Brazil Emerging

This volume is a critical inquiry into the social project and socioeconomic realities of emerging Brazil, a country that faces profound changes. A team of acknowledged specialists on Brazil's complex configuration addresses state policies, social dynamics and economic constraints and opportunities for emancipation. Chapters adopt long-run perspectives on the development of the Brazilian welfare state, limits and opportunities for emancipation in the labor market, the scope and depth of social policies such as "Bolsa Família" and Rio's Peacemaking Police Units (UPP), social movements—in particular, the Movement of the Landless (MST)—cultural policies at the federal level, the role of media in the country's democratization project, and how two important commodities (sugar and oil) shape the identities of blacks and whites in Bahia. This book is essential reading for all those interested in understanding what kind of Brazil has acquired a prominent global position and what hurdles it faces to consolidate its position as a global player.

Jan Nederveen Pieterse is Mellichamp Professor of Global Studies and Sociology at University of California, Santa Barbara.

Adalberto Cardoso is Professor of Sociology at IESP-UERJ.

Routledge Studies in Emerging Societies

SERIES EDITOR: JAN NEDERVEEN PIETERSE,
University of California, Santa Barbara

The baton of driving the world economy is passing to emerging economies. This is not just an economic change, but a social change, with migration flows changing direction towards surplus economies; a political change, as in the shift from the G7 to G20; and over time, cultural changes. This also means that the problems of emerging societies will increasingly become world problems. This series addresses the growing importance of BRIC (Brazil Russia India China) and rising societies such as South Korea, Taiwan, Singapore, Indonesia, South Africa, Turkey, the UAE and Mexico. It focuses on problems generated by emergence, such as social inequality, cultural change, media, ethnic and religious strife, ecological constraints, relations with advanced and developing societies, and new regionalism, with a particular interest in addressing debates and social reflexivity in emerging societies.

1 **Global Modernity, Development, and Contemporary Civilization**
Towards a Renewal of Critical Theory
José Maurício Domingues

2 **Globalization and Development in East Asia**
Edited by Jan Nederveen Pieterse and Jongtae Kim

3 **Brazil Emerging**
Inequality and Emancipation
Edited by Jan Nederveen Pieterse and Adalberto Cardoso

Brazil Emerging

Inequality and Emancipation

**Edited by Jan Nederveen Pieterse
and Adalberto Cardoso**

LONDON AND NEW YORK

First published 2014
by Routledge
711 Third Avenue, New York, NY 10017

Simultaneously published in the UK
by Routledge
2 Park Square, Milton Park, Abingdon, Oxfordshire OX14 4RN

*Routledge is an imprint of the Taylor and Francis Group,
an informa business*

First issued in paperback 2015

Cover photo: A boy at the favela of Cantagalo, Rio de Janeiro, facing Ipanema. Photographer: Ricardo Azoury.

The right of Jan Nederveen Pieterse and Adalberto Cardoso to be identified as the authors of the editorial material, and of the authors for their individual chapters, has been asserted in accordance with sections 77 and 78 of the Copyright, Designs and Patents Act 1988.

Library of Congress Cataloging-in-Publication Data
    Brazil emerging : inequality and emancipation / Edited by Jan Nederveen Pieterse and Adalberto Cardoso.
        pages cm. — (Routledge studies in emerging societies ; 3)
    Includes bibliographical references and index.
    1.   Income distribution—Brazil. 2.   Brazil—Economic conditions—21st century. 3.   Brazil—Economic policy. 4.   Brazil—Social conditions—1985-   I.   Nederveen Pieterse, Jan editor of compilation.
    II. Cardoso, Adalberto Moreira editor of compilation.
    HC190.I5B73 2013
    330.981—dc23
    2013005296

ISBN 978-0-415-83704-0 (hbk)
ISBN 978-1-138-95291-1 (pbk)
ISBN 978-0-203-48213-1 (ebk)

Typeset in Sabon
by IBT Global.

Contents

Figures

Graphs

Tables

Acknowledgments

Most chapters in this book were originally presented as keynote papers at the fourth Global Studies conference held in Rio de Janeiro in July 2011. We thank the participants in this conference and, in particular, the staff of Common Ground for producing the conference and dealing with practicalities. We also thank keynote speakers whose contributions could not, for one reason or another, be included in this volume, including Elisa Reis, Marcelo Córtes Neri, and João Paulo Rodrigues Chaves of MST, São Paulo.

Chapters by Jan Nederveen Pieterse and Gustavo Lins Ribeiro were published previously in earlier versions (respectively, in Rudolf Traub-Merz, ed. *Growth through Redistribution?: Income Inequality and Economic Recovery*, Friedrich-Ebert-Stiftung, Shanghai Coordination Office for International Cooperation, 2012, and in *Postcolonial Studies*, 14, 3, 2011) and we gratefully acknowledge permission to publish them in this volume.

Introduction

Jan Nederveen Pieterse
and Adalberto Cardoso

> From our perspective as emerging economies, the resources that are
> needed to overcome hunger and poverty may be considerable but are
> quite modest when compared to the cost of rescuing failed banks and
> financial institutions that are victims of their own speculative greed.
> (Lula da Silva 2010: 22)

All eyes are now on emerging markets, and in emerging economies all
eyes are on the middle class, which is growing by 70 million a year. The
middle class matters economically as consumers and culturally and politically because in values and lifestyle they seem to be similar to the middle
class in the West. Focusing on the middle class comes natural to mainstream sensibilities and in the process makes the 'rise of the rest' appear as
an extension of the rise of the West. However, whereas the growing middle
class is important, the central challenge of modernities, old and new, is
whether they are able to integrate the poor. This involves structural reforms
(land reform, tax reform, investments in the countryside, education and
social policies), cultural legacies and perspectives on social inequality, and
agency and social movements. Arguably, how and on what terms the poor
majority participates in the new modernities determines whether the new
modernities will be viable and sustainable twenty years hence. Whereas
the rise of emerging societies is a given and global multipolarity is no longer in question, what is in question is how the rise of emerging economies
affects the rural and urban poor and the informal sector. Is emancipation
of the poor part of the horizon—culturally, socially, and in terms of the
economic growth model—of Brazil, Latin America, and other emerging
societies? In Brazil the poor are 20 percent of the population and 6 percent
are extremely poor. This volume is a critical inquiry into the social project
and socioeconomic realities of emerging Brazil.

As one of the BRICs (Brazil, Russia, India, China), Brazil ranks as a
high-promise economy. Like other BRICs Brazil is an emerging economy of
almost continental size, with multiple time zones and climate zones, from
the temperate south to the tropical north. One view holds that "in some
ways, Brazil outclasses the other BRICs. Unlike China it is a democracy.

Unlike India, it has no insurgents, no ethnic and religious conflicts nor hostile neighbors. Unlike Russia, it exports more than oil and arms and treats foreign investors with respect" (*The Economist* 2009). Be that as it may, as with the other BRICs, high promise doesn't refer to the whole economy but to specific sectors—in Brazil's case, agro-mineral exports (notably soy, meat, iron ore, and petrol). The BRIC is a deceptive category. Whereas the units are countries—Brazil, Russia, India, China—the interest is focused on specific economic sectors that are deemed strategic and promise high returns, which is crucial for institutional investors in the West seeking high returns. For development studies and global sociology, the point is to ground the blue skies of the BRICs in social realities, substituting the notion of emerging markets for emerging societies (Nederveen Pieterse and Rehbein 2012) and focusing not merely on growth but on the quality of growth, on social inequality, and on the frontiers of emancipation.

According to Rubens Ricúpero (former secretary general of UNCTAD), there are "four reasons to believe in Brazil": commodities—the commodities boom driven by emerging markets, which for Brazil ranges from iron ore to agricultural products; petrol—the country is poised to become a medium-size net oil exporter; demography—the ratio of dependents to people of working age has dropped sharply since the 1990s to 48:100, which enables a greater quality of social spending; and urbanization—in that the worst problems of frantic and chaotic urbanization are behind it (*The Economist* 2010). In addition, Brazil is at peace with all ten of its neighbors and is a stable democracy.

Brazil is a fast-rising agro-mineral and industrial exporter with a social structure shaped by slavery, landlords with large holdings, and a highly concentrated economic structure. The country has long been associated with extreme inequality. In the 1970s Brazil was the international example of 'immiserizing growth', growth that pauperizes the majority of the population. 'Brazilianization' became international shorthand for steeply growing social inequality. It earned Brazil the nickname Belindia, somehow a combination of Belgium and India (which are combinations themselves). Brazil still ranks among the societies with very high inequality, with a poverty level of 20 percent (40 million) in 2009. Also among BRIC countries Brazil has the highest inequality. Comparing the proportion of national income received by the top 10 percent of the population to that received by the bottom 10 percent among the BRICs in 2008, China comes out the least unequal with the bottom 10 percent earning 3.5 percent and the top 10 percent earning 15 percent, whereas proportions in the other three are closer together, and Brazil shows the widest social inequality (Central Intelligence Agency 2010) (Table 0.1).

There have been major changes and a significant reduction of poverty and inequality since the governments of F. H. Cardoso, Lula da Silva, and Dilma Rousseff. Brazil's Gini coefficient has fallen more than five points since 2000, to 0.55 (*The Economist* 2011). According to World Bank data, Brazil's Gini index was 54.52 in 2008 and 51.9 in 2012.

Table 0.1   The Proportion of National Income Received by the Top 10 Percent and Bottom 10 Percent, 2008

BRIC	Top 10%	Bottom 10%
China	15%	3.5%
Russia	30.4%	1.9%
India	31.1%	3.6%
Brazil	43%	1.1%

According to economist Marcelo Córtes Neri, "Brazilian income inequality has been falling steadily since (and only after) the very beginning of the last decade. Between 2001 and 2009 per capita incomes of the 10% richest grew at 1.49% per year while the incomes of the 10% poorest grew at 6.79% per year. . . . Brazil is about to reach the lowest income inequality level since 1960, when the records started" (Neri 2010: 9, 11). "From 2003 to 2009, real per capita income, according to the PNAD [National Household Sample Survey] was 69% for the poorest people (and 12.6% for the 10% wealthiest people) . . . the growth rate in the poorest bracket was 550% higher than the growth of the wealthiest, which is what we call the spectacle of growth, but only for the poor, who have been experiencing Chinese growth rates" (Neri 2011: 14). According to UNDP data, Brazil's Human Development Index rose steeply from 2000 (Figure 0.1).

The Lula government increased the minimum wage and extended the Bolsa Família program that was initiated under the Cardoso government.

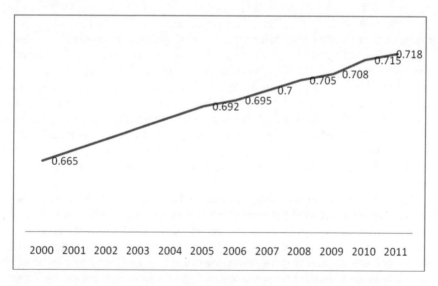

Figure 0.1   Human Development Index Brazil. Source: http://hdr.undp.org.

The Bolsa Família program provides aid to families living below the poverty line, conditional on recipients meeting medical and educational requirements for their children. Currently some 49 million people (a quarter of the population) receive such stipends. The benefits of the program have been criticized (Petras 2009) and Bolsa Família (BF) has generated wide international discussion, also as part of the general policy of conditional cash transfers in developing countries (Lomeli 2008; Soares 2011; Bastagli 2011). Cash transfers are supposed to invest state funds towards human development gains for the most disadvantaged, but do they transform people's life circumstances and contribute to social emancipation?

The first two chapters in this volume, by Sonia Fleury and Marcelo Medeiros et al., address Brazil's social policies and consider their impact in reducing social inequality as well as their limitations. Sonia Fleury views Brazil's efforts at building a welfare state as part of Brazil's process of building a democracy.

Bolsa Família shares the usual limitations of social programs that are unconnected to growth policies and leave major policy areas untouched, such as land reform and reform of the tax system. In this volume Nederveen Pieterse offers a panorama of the relationship between economic growth and social policies in emerging societies and developing countries generally and discusses the case for inclusive development.

These trends are significant well beyond Brazil. The major challenge facing emerging societies is to achieve shared growth, a growth pattern that benefits the majority of the population. If Brazil can do it, especially Brazil, the erstwhile poster child of pauperizing growth, others can do it. The key questions are how has this remarkable reduction of poverty been achieved and is it sustainable. According to Marcelo Neri, the major elements have been the Real plan of the Cardoso government, job growth, in particular the growth of formal employment, and social policies, in particular Bolsa Família. According to Neri, "labor earnings explain two thirds (66.86%) of the total inequality reduction between 2001 and 2008, next come the contribution of social programs with emphasis on the Bolsa Família program (Family Grant) and the Bolsa Escola (School Grant) . . . that explain 17% of the higher equality obtained while social security benefits explain 15.72%" (Neri 2010: 17). In other words: two-thirds job and wage growth and one-third social programs. The 17 percent effect of social programs may be small in the aggregate but is profoundly significant because it affects the very poor.

The Real plan has been significant in bringing down inflation. Job growth and expansion of the formal labor market are contentious and generate wide discussion. A 2010 report noted, "Brazil's rise now is being sustained by big shifts in the domestic economy and in particular by the rise of new urban middle and working class consumers" (*The Economist* 2010; cf. Neri 2012). Besides the Lula government's minimum wage policy, its Zero Hunger platform, and the extension of Bolsa Família, "there has also been a big increase in the number of formal jobs, with the annual rate of new positions doubling over the past six years" (Lapper 2010). In this volume Adalberto Cardoso examines Brazil's labor market, as the central and most important gateway of emancipation. Cardoso notes the size and persistence of informal labor, which for many in the workforce is a dead end.

Brazil's recent growth spurt relies to a large extent on the commodities demand of China and other Asian drivers (Rohter 2010; Oliveira 2012). In 2009 China became Brazil's biggest trading partner, replacing the US. It yields Brazil's profile as an 'agricultural superpower', a leading exporter of soy, beef, chicken, and grain, besides iron ore and petrol. But an economy based on commodities exports is narrow and bust follows boom, even if demand is likely to be sustained over a longer period than previous commodities cycles because it is driven by long-term trends of population growth and urbanization in Asia (Sinnott and de la Torre 2010; *The Economist* 2010). Signs of a boom economy are abundant. São Paulo sets "records for conspicuous consumption" such as "having one of the largest Ferrari dealerships in the world, a large fleet of private jets and being the only city in the world to boast four Tiffany stores" (Ramsey 2009); salaries of executives are higher than in London (Naím 2011), and office space is more expensive than in Manhattan (as is also the case in Beijing).

Whether these trends are sustainable is in question. The share of manufacturing in Brazil's total economic production stood at 19.2 percent in 2004 and declined to 14.2 percent in 2011, a trend that some analysts have called 'precocious specialization' (Oreiro and Feijó 2010) or 'regressive specialization', with the loss of strategic links in important production chains. The industrial sector is now notable mostly for a few global players such as aircraft (Embraer), petrol (Petrobrás), automobile assembly, and light industries, alongside a growing services sector. Others see signs of a Brazilian kind of 'Dutch disease', that is, deindustrialization due to currency appreciation resulting from specialization in commodity exports (Souza 2009; Bresser-Pereira 2010).

Other factors discussed in this context are the after-crisis policies of low interest rates in the US and Europe, in particular the Federal Reserve policy of quantitative easing, which produces, in Dilma Rousseff's words, an "immense, fantastic, extraordinary sea of liquidity that finds its way to our economies in search of returns that it can't find in its own" (quoted in Lyons 2011). In developing countries the capital inflows produce inflation and rising prices of food and real estate. The first six months of 2011 saw a 260 percent spike in foreign direct investment in Brazil. "Ultra-low interest rates in the west continue to flood faster-growing markets such as Brazil's with capital, pushing their currencies higher" and "households spend about a quarter of their disposable income servicing their debt" (*Financial Times* 2011). Brazil has seen the Real appreciate, interest rates rise (nominally 12.5 percent, in real terms near 5 percent), and economic growth slow. In response to what is deemed a 'currency war', the government has put in place a series of measures including a tax on incoming foreign capital, currency intervention, subsidized loan programs, Buy Brazil policies, and interest rate reduction (below 8 percent nominal, 2 percent real in 2012).

Cheap Chinese manufactures compete with domestic industries, especially when the Real is high. "For some years after 2000 Brazil had a trade surplus with China, but this trend was reversed in 2007 when Brazil's imports from China reached $12.6 billion. These are mainly electronic (media and high technology) and chemical products, related to the global production strategies of TNCs" (Fernández Jilberto and Hogenboom 2012: 13). In addition, "some Brazilian manufacturers have begun migrating production to China or India," particularly in industries such as automotive components (Leahy 2011). Competition with Brazil's manufactures exports in third markets is also a consideration.

Brazil has seen major social movements: a strong abolitionist movement; strong labor; indigenous, environmental, and women's movements; MST; and anticorruption movements. Brazil is the home of trendsetting forms of participatory democracy such as Porto Alegre's participatory budget (Burity 2009; Seisdesos 2010). Brazilian social movements have initiated and host the World Social Forum in Porto Alegre. These movements form the backdrop to the rise of the PT and the Lula and Rousseff governments and the social commitment that inspires the Bolsa Família and other social reforms.

Two chapters in this volume examine the role of social movements in Brazil. Ilse Scherer-Warren offers a general overview and analysis, and Breno Bringel focuses on MST, the movement of the landless, Brazil's largest and most significant social movement. After major rural-urban migration, now only 11 percent of Brazil's population lives in the countryside, so arguably land reform is no longer a major issue. Also for MST, as discussed by Bringel, the agenda has shifted to the problems faced by the urban poor.

Another dimension of the integration of Brazil's poor in the folds of society is racism affecting Afro-Brazilians (Telles 2004). The problem with racism in Brazil is that it has been denied for many decades by public policy makers, cultural elites, and the white population at large. As classically noted by Oracy Nogueira (1955), social prejudices in Brazil are based on a person's skin color, not her ethnicity (ethnic origin and identity). Racial discrimination shows, among other things, in the fact that nonwhites are 55 percent of the population but seven in ten of the poorest Brazilians are nonwhite. Thus inclusive development in Brazil also means opening ways of social mobility and emancipation for black Brazilians. In this volume Livio Sansone reflects on the transformation of identities in Bahia-Salvador.

In spite of the dynamism of its social movements and the increasing quality of its democracy (see Carolina Matos on the relation between media and democracy in this volume), Brazil faces several hurdles. Its governance is notoriously complex, a matter of complex coalitions with entrenched interests and elites, old and new.

The educational system is a major mechanism for the reproduction of economic and social inequalities. Brazil has universalized elementary schooling (eight years), but the quality of public schools is poor (Ribeiro 2009; Barbosa 2009). Wealthy families can pass on their cultural capital to their heirs via high-quality private elementary and secondary schools that open the doors to quality public higher education, which members of poorer families can seldom reach. Affirmative action policies have reduced the entry barriers to Afro-descendants in higher education, but this applies to only a minority. The poor and Afro-descendants are selected out of a virtuous educational track mostly during the transition from elementary to secondary education (Torche and Ribeiro 2012), and their effective inclusion in social and economic dynamics is contingent on the elimination of these early barriers to school achievement.

Social violence related to drug and arms trafficking is another major problem in urban Brazil. Poor areas in most of the country's metropolises are controlled by competing factions of violent, heavily armed organized groups and gangs. Even if incarcerated, gang leaders continue to act from within penitentiaries, where they are safeguarded from peer violence. All medium-size and large cities suffer from drug-related armed crime. More recently, crack consumption (and the violence that comes with it) has soared and is considered epidemic by the Ministry of Health. In this respect, the Pacifying Police Units (UPP) experiment in Rio de Janeiro is proving effective in regaining control over territories once dominated by armed gangs, reducing armed violence (especially homicides), but the impact on actual drug consumption and trafficking is still in question (Glenny 2012). In this volume Erica Mesker discusses the UPP operations and notes the growing gap between pacified and nonpacified favelas.

Unlike some of the BRICs, Brazil is culturally integrated by one language. After decades of public cultural policies that tried to frame a homogeneous national identity, the government has recently begun to promote diversity and difference in the cultural sphere, including affirmative action for Afro-Brazilians, women, indigenous populations, and other minorities. National identity is increasingly associated with the idea of a creative cultural melting pot in which tradition and innovation mix and are equally valued. In this volume Myrian Santos discusses the evolution of Brazil's cultural policies.

Among the BRICs Brazil is a historical newcomer to international influence, and at times its inclusion among the BRICs is questioned (Armijo and Burges 2010). Nevertheless, the idea of a 'Brasília consensus' is in the air, and Brazil's policy of pragmatic 'cooperative multipolarity' has contributed to several international initiatives in which Brazil has acted as a counterweight to the West, such as stepping out of the Doha round of the WTO, the G20, reforms of the IMF to empower emerging economies in its decision making, and cooperation with Cuba and Iran. In the recent economic crisis, Brazil has loaned the IMF $20 billion to help rescue policies in

Europe. Part of Brazil's 'rainbow' strategy is that it has more embassies in Africa than the UK (Leahy 2012). Whereas Brazil's 'tropical modernism' is widely acknowledged, in this volume Gustavo Lins Ribeiro takes a further step and reflects on Brazil's cosmopolitanism, that is, tropical cosmopolitics as a contribution to transnational emancipation.

The agenda of emancipation of the political coalition in office since 2003, headed by the Workers' Party with its historical roots in social and labor movements, can be summed up in the catchword 'neo-developmentalism'. At the microeconomic level the agenda combines state investments in infrastructure (directly or through public-private partnerships), stimulating private sector investments in manufacturing (mostly via subsidized loans of the National Bank of Social and Economic Development, BNDES, but also via tax reductions), subsidies for innovative research and development, attracting productive foreign investment, and many other measures. At the macroeconomic front there are slight shifts from the neoliberal framework: control of the exchange rate to protect national industry, lower interest rates (to stimulate investments and credit to the families), and some tolerance of inflation, combined with pro-cyclical fiscal austerity, that is, state spending during crisis and contraction during growth. At the social level, social policies redistribute income, reduce inequality and poverty, favor upward social mobility, and in the process consolidate the domestic market as the chief mechanism of sustained growth.

This project is complex and ambitious, and its consequences in the medium and long run are not clear for they will depend on how it intertwines with the social and economic fabric, where a few very powerful agents (in the private sector and the state) have virtual veto power in strategic areas such as the mass media and the financial sector. The mass media in Brazil are highly concentrated and their leading spokespersons are clearly anti-state and pro-finance. This means that the emancipation agenda must find its way in an inhospitable ideological environment and a very competitive political system that raises the costs of mistaken decisions by incumbents of power.

REFERENCES

Armijo, L. E., and S. W. Burges. 2010. "Brazil, the Entrepreneurial and Democratic BRIC." *Polity* 42 (1): 14–37.

Barbosa, Maria L. 2009. *Desigualdade e desempenho: uma introdução à sociologia da escola brasileira*. Belo Horizonte: Argumentum.

Bastagli, F. 2011. "Conditional Cash Transfers as a Tool of Social Policy." *Economic and Political Weekly* 46 (21): 61–66.

Bresser-Pereira, Luis C. 2010. *Globalization and Competition: Why Some Emergent Countries Succeed While Others Fall Behind*. Cambridge: Cambridge University Press.

Burity, Joanildo A. 2012. "Inequality, Culture and Globalization in Emerging Societies: The Brazilian Case." In *Globalization and Emerging Societies:*

Development and Inequality, edited by J. Nederveen Pieterse and Boike Rehbein, 161–81. London: Palgrave Macmillan.

Central Intelligence Agency. 2010. *The World Factbook*. https://www.cia.gov/library/publications/the-world-factbook/index.html (accessed April 3 2012).

The Economist. 2009. "Brazil Takes Off." November 12.

———. 2010. "Four Reasons to Believe in Brazil." July 26.

———. 2011. "Inequality Unbottled, Gini Inequality Is Rising." January 20.

Fernández Jilberto, Alex E., and Barbara Hogenboom. 2012. "Latin America and China: South-South Relations in a New Era." In *Latin America Facing China: South-South Relations beyond the Washington Consensus*, edited by Alex E. Fernández Jilberto and Barbara Hogenboom, 1–32. Oxford: Berghahn (original published 2010).

Financial Times. 2011. "Brazil's Currency War Wounds." [Comment]. July 8, 8.

Glenny, Misha. 2012. "The Fight for Rio's Favelas." *Financial Times*, November 3–4, 1–2.

Lapper, R. 2010. "Evidence of New Self-Confidence Is Everywhere." *Financial Times*, November 15, 6.

Leahy, Joe. 2011. "Brazilian Factories Tested by Chinese Imports." *Financial Times*, January 31, 6.

———. 2012. "The Brazilian Economy." *Financial Times*, January 11, 7.

Lomeli, E. V. 2008. "Conditional Cash Transfers as Social Policy in Latin America: An Assessment of Their Contributions and Limitations." *Annual Review of Sociology* 26 (1): 58–74.

Lula de Silva, Luis Inácio. 2010. "The BRICs Come of Global Age." *New Perspectives Quarterly* (Summer): 21–22.

Lyons, J. 2011. "Dark Side of Brazil's Rise." *Wall Street Journal*, September 10, A1–14.

Naím, M. 2011. "End the Party before Brazil's Bubble Bursts." *Financial Times*, June 1, 10.

Nederveen Pieterse, J., and Boike Rehbein, eds. 2012. *Globalization and Emerging Societies: Development and Inequality*. London: Palgrave Macmillan.

Neri, Marcelo Córtes. 2010. *The New Middle Class in Brazil: The Bright Side of the Poor*. Rio de Janeiro: Fundaçao Getulio Vargas.

———. 2011. *Income Inequality on the Decade in Brazil: Selected Articles*. Rio de Janeiro: CPS/FGV.

———. 2012. *A Nova Classe Média*. Rio de Janeiro: CPS/FGV.

Nogueira, Oracy. 1955 [1998]. *Preconceito de marca*. São Paulo: EDUSP.

Oliveira, H. A. de. 2012. "Brazil and China: from South-South cooperation to competition." In *Latin America Facing China: South-South Relations beyond the Washington Consensus*, edited by Alex E. Fernández Jilberto and Barbara Hogenboom, 33–54. Oxford: Berghahn (original work published 2010).

Oreiro, José L., and C. A. Feijó. 2010. "Desindustrialização: conceituação, causas, efeitos e o caso brasileiro." *Revista de Economia Política* 30 (2): 219–32.

Petras, James. 2009. *Global Depression and Regional Wars*. Atlanta, GA: Clarity Press.

Ramsey, J. 2009. "Undisputed Number One." *Financial Times*, November 17, 4.

Ribeiro, Carlos A. 2009. *Desigualdade de Oportunidades no Brasil*. Belo Horizonte: Argumentum.

Rohter, L. 2010. *Brazil on the Rise: The Story of a Country Transformed*. New York: Palgrave.

Rousseff, Dilma. 2011. "Brazil Will Fight Back against the Currency Manipulators." *Financial Times*, September 22, 11.

Seisdedos, Paul C. 2010. "'Late neoliberalism' in Brazil: social and economic impacts of trade and financial liberalization." In *Confronting Global Neoliberalism:*

Third World Resistance and Development Strategies, edited by R. Westra, 39–66. Atlanta, GA: Clarity Press.

Sinnott, E., J. Nash, and A. de la Torre. 2010. *Natural Resources in Latin America and the Caribbean: Beyond Booms and Busts?* Washington, DC: World Bank.

Soares, Fario Veras. 2011. "Brazil's Bolsa Família: A Review." *Economic and Political Weekly* 46 (21): 55–60.

Souza, Cristiano R. S. 2009. "O Brasil pegou a doença holandesa?" PhD diss., University of São Paulo.

Telles, Edward E. 2004. *Race in Another America: The Significance of Skin Color in Brazil.* Princeton, NJ: Princeton University Press.

Torche, Florencia, and Carlos C. Ribeiro. 2012. "Parental Wealth and Children's Outcomes over the Life Course in Brazil: A Propensity Score Matching Analysis." *Research in Social Stratification and Mobility* 30(1): 79–96.

1 Building Democracy in an Emerging Society
Challenges of the Welfare State in Brazil

Sonia Fleury

This chapter analyzes some topics related to the challenges of building democracy and social inclusion in Brazil, one of the most unequal societies in the world. The focus will be on the efforts of building up a welfare state (WS) in extremely adverse conditions.

Recently, Latin American countries had to face the challenges of improving social inclusion and economic redistribution, as well as consolidating democratic institutions. Facing the pressure for social inclusion was all the more difficult in Latin American countries because these were societies characterized by some of the worst income disparities in the world, high degrees of labor market informality, and the presence of powerful actors with vested interests in maintaining the old social protection system, which was stratified and exclusive. These stratified and centralized protection models, restricted to workers employed in the formal labor market, were reformed in the last quarter of the twentieth century, because they were seen as far from egalitarian and as actually reproducing social injustice. These reforms engendered new designs of social policies in a double movement of universalizing social coverage as well as targeting poverty and urban violence.

In addition to the challenge of increasing social inclusion and fighting poverty, the construction of new democratic and inclusive institutions had its takeoff in a context of macroeconomic adjustment aimed at combating hyperinflation, in which there was pressure from international agencies to downsize the state and reduce its power. Moreover, this new global pattern of productive, economic, and social development was characterized by Kim (2009: 18) as postindustrialization, postclassism, and postfamilism, indicating that in the last quarter of the twentieth century the basic conditions supporting the welfare state were eliminated, because economic growth was no longer based on manufactured production, social mobilization was disconnected from class struggle and organized labor, and the modern family has deviated from a male breadwinner model thus reflecting a new gender and age structure.

In the last two decades, Brazil has been trying to build a welfare system based on the principle of extending social rights to all citizens through universal social policies, a principle that is enshrined in the social security concept of the Federal Constitution of 1988. The importance of studying the Brazilian experience lies not only in observing the path dependency impact of the former institutional social protection model but also in addressing the absence in Brazil of the most important requirements pointed out by the literature as necessary for the successful development of an egalitarian welfare system. This chapter reviews the WS literature and compares it to the Brazilian experience in order to highlight this discrepancy and analyze its effects on the emerging system. It contends that the statecraft of the institutional framework of social policies that occurred during the 1990s had to cope with dreadfully restrictive economic constraints. The analysis of the way the economic, cultural, and institutional constraints have affected the construction of the Brazilian welfare state (BWS) demonstrates how the absence of important requirements for WS development has led to a mixed and sometimes hidden institutional framework that enables the prevalence of private interests in public policy.

The reduction of poverty is an important outcome of recent social policies as well as of continuous increases in the minimum wage. Nonetheless, government has not prioritized the universal systems of health and education, which increases the stress and contradictions between the demands of the citizens and the state's capacity to provide both access and high-quality services. The precariousness of social inclusion, both in the market and in the community of citizens, generates demands for a comprehensive public sphere and new institutional arrangements of a deliberative democracy capable of generating conditions not only to guarantee social rights but also to lead to emancipation.

The European bias in the WS literature leads to an implicit conclusion that this experience cannot be replicated in other contexts, in which some of the presumed preconditions are absent. Even the maintenance of the WS in these original countries seems to be threatened by the profound changes in the economic and social pillars of the social democratic construct. This concern is based on the decline of labor-based societies and the replacement of social cohesion bonds by negative competitive individualism (Castel 1995). The reduction of the middle class and the growth of inequalities magnified the income gap between the extremes of the social scale. The difficulty of enforcing an equitable tax system in a predominantly financial economy and the pressure to prevent financial bankruptcy led to cuts in public expenditure, productive investments, and redistributive policies.

Beyond that, the importance of economic interests in social sectors is expanding, with powerful profit-seeking actors like pharmaceutical industries, insurance companies, and private service-providers now playing a dominant role in public polices by influencing their design and management as well as by shaping the consumption of social services. The analysis of the opportunities for building up social protection in a region such as Latin America is overdue, considering the enduring and pervasive high levels of inequality extant there, in spite of the influence of the European social ideals in these countries' intellectual elite. The recent wave of democracy in Latin America put social protection and poverty reduction in the public agenda, notwithstanding the legacy of authoritarian regimes, the preponderance of the elite's vested interests in the public arena, and the dispute for scarce public resources in a context of economic crisis.

The global economic crisis is now narrowing the distance between developed welfare societies and laggard ones. Therefore, it is necessary to revisit European WS theories in order to establish a dialogue with experiences in adverse contexts.

THE LIMITS OF WELFARE STATE THEORY TO EXPLAIN THE EXPERIENCE OF EMERGING SOCIETIES

Most studies on the WS derive their conclusions from comparing European experiences. This literature has pointed out some conditions associated with the emergence and consolidation of the WS. The starting point is considered to be the industrialization process (Rimlinger 1971), along with the emergence of social insecurity in a context of urbanization. The roots of the WS development are based, on one hand, on the fracture of the traditional communitarian bonds of solidarity and their displacement by class organization and new identities. On the other hand, they are embedded in the development of civil service and state administrative structures. In this sense, the WS was responsible for forging new bonds in complex societies and also for creating a new institutional pattern of redistributive conflict resolution, launching the basis for social cohesion and integration.

The WS and the competitive political party system are seen by Offe (1984) as the main institutions capable of promoting the coexistence of capitalism and democracy in its virtuous cycle. Contradictorily, at the end of this cycle of economic growth in a regulated market, both features were regarded as inflexible obstacles for renewing the capitalist economy. The convergence perspective is expressed by Wilensky's (1975: 27) conclusion that, in spite of ideological and political regimes, the WS is the most persistent structural tendency in the development of modern societies. In this manner, he associated the WS with economic development and the social modernization process.

Differently from this *nomothetic* (Takegawa 2009) perspective, other authors have highlighted the importance of path dependency in the process, stressing that demands for the emergence of social protection have received different responses according to the previous institutional, political. and economic environment. Flora and Alber (1981) identify the emergence of the universal model of social protection as dependent on the strength of a homogeneous working class in fighting for social rights, as well as on the capacity of each society to institutionalize this conflict through democratic procedures.

Starting with Titmuss's (1958) classical typology of social services, we can find a confluence of criteria encompassing both the relations between market and public sector and the degree of redistribution through social policies. His typology is based on the extension of social rights and on the double movement of expanding state structures and policies on one hand, and redistributing resources according to necessities on the other hand. According to Wincott (2011: 358), both Titmuss and Marshall were late converts to the use of the term 'welfare state', a name actually used by the critics of social policies in England. Marshall (1967) emphasizes the core role of citizenship status in the coexistence of an egalitarian political principle within a class-based economy. In his classic article on social classes and citizenship, he identified three kinds of rights as components of citizenship—civil, political, and social—each having its own path and institutional structure.

The WS is considered a new arrangement for consensus building and to convey conflicts to the redistributive arena, where they can have a negotiated solution. Different patterns of social protection were identified according to where the main aim of inclusion was centered: on the poor and other vulnerable groups, on labor fractions of the workforce, or on the citizens. The inclusion of social rights as part of the status of citizenship represented the most paradoxical solution for the distributive conflict in a class economy, because it had generated a public sphere not primarily subordinated to the process of accumulation, an anti-value mechanism according to Oliveira (1988). Esping-Andersen (1990) shed new light on social rights, which were reconceptualized in terms of their degree of 'de-commodification', meaning the capacity to take social reproduction away from merchandise circulation, producing a new, socially stratified blueprint. He also underlines the importance of institutionalization when he analyzes the consequences of the WS crisis at the end of the last century in three different regimes—the liberal welfare regime, the corporatist-state regime, and the social democratic regime. His analysis concludes that the political and institutional mechanisms of interest representation and consensus building, especially in the social democratic arrangement, have a strong impact on the preservation of employment and social rights (Esping-Andersen 1995: 77).

To summarize, according to the WS literature the development of the WS has been associated with an array of elements related to the progress of the capitalist economy, as well as with the transformation of societies as a consequence of urbanization and industrialization processes that imposed a new division of labor and led to more complex social relationships. The WS is also part of the state-building process, and this expansion is a component of the democratization of power and wealth in a mass society, with the emergence of new collective actors and political organizations. Nonetheless, it contributed to the creation of a more cohesive society, based on social principles of solidarity, in which social inclusion was widespread. However, during the crisis of the capitalist economy that started in the

1970s, the institutional mechanisms of social protection were accused of impeding the renovation of the productive relations in order to increase competition and productivity. Lately, the loss of affiliation links was pointed out as being responsible for the crisis of sociability, the rise of insecurity, the sprout of negative individualism, and the replacement of policies of inclusion by policies of labor-market insertion—also called active inclusion or workfare regime (Castel 1995).

Recent studies on the WS in developing societies have pointed out the incapacity of established theories to explain the efforts to build up social protection in different contexts in developing societies. To rethink social policy in developing countries, Wood and Gough (2008: 313) suggested that Esping-Andersen's original typology of welfare state regimes must be seen according to the degree of in/security and in/formality of rights and correlatives duties in each society. To this effect they added two mixed variants to the previous scheme: the liberal-informal WS regimes in Latin America and the productivist WS in Asia, stressing the informality of benefits and programs in the first case and the role of employers in the second.

Regardless of recent efforts to conceptualize and understand what is happening beyond Europe, the interpretation of the Asian late-coming WS is criticized for being unable to include some important variables such as the influence of international circumstances on these countries (Takegawa 2009: 81). Takegawa hypothesizes that whereas domestic factors determine the time of the takeoff, international circumstances govern the development of the WS. The takeoff of the WS in Japan in the mid-1970s was embedded in the liberalism, because it occurred during the crisis of the traditional WS. These constraints led to a relatively low level of public social expenditure, what Takegawa (2009: 89) called *universalism rationing*, that is, the simultaneous adoption of pro-welfare and anti-burden policies. The result was the mixed Japanese system, involving families and enterprises in social protection. However, this productivist model was not a consequence of Confucianism. Instead, this mix is the product of expanding welfare protection with low public social expenditure and strong state intervention in the economy.

Kim (2009) also denies the application of the concept of hybrid welfare to explain the South Korean reform, because this kind of label is based on value judgments about welfare states' leaders or laggards, instead of trying to understand the peculiar dynamics of each reform. In South Korea a peculiar process of controlling and restructuring of the WS occurred simultaneously with its formation. This process of going backward and forward in the universal WS project can be understood only if one combines the need for social inclusion after the fall of the military regimes along with the specific governmental ideologies in a context of economic constraints.

Although the process of Latin American WS building is older than many others in Europe, it has similarities with the Asian cases, because the international economic context had the same effect of constraining the expansion of social expenditures, creating an obstacle to the promotion of social inclusion in the new democratic regimes. On the other hand, the former institutional framework for corporative, stratified, and segmented protection gave rise to a community policy with strong veto power over the reforms of these systems as well as the existence of a market for the provision of social services. The incapacity of weakened nation-states to build consensus regarding the reform guidelines and to regulate the market were some of the regional challenges. Barrientos (2008) characterized the changing WS in Latin American during the 1990s as a conversion of the conservative-informal welfare regime into a liberal-informal regime. He labeled the previous Latin American regime as conservative-informal to identify the stratified nature of social protection, and concluded that currently there is a change in this pattern towards a liberal-informal welfare regime as a consequence of market deregulation. The new regime is also characterized as informal because the replacement of collective insurance for individual savings and free-market provision of services did not extend formal rights to social protection for those employed in the informal labor market. Nevertheless, the category liberal-informal is unable to differentiate among the reforms that occurred in the last quarter of the twentieth century; therefore it is not a useful tool for explaining variations in the WS in this region.

Although the reforms occurred in a period of prevalence of liberal ideology and economic adjustment, there were differentiated institutional solutions for the national explosion of social demands for inclusion. The strong global trend towards state contraction and liberalization of labor protection obscures the originality of the regional reforms as well as the plurality of WS reforms in the region.

In order to account for this plurality and originality, it is useful to review the literature on WS building in the region and also on the models of reforms after the 1980s. The leading comparative study on social policy development in Latin America was made by Mesa-Lago (1978). Comparing the development of the retirement and pension systems, he adopts the timing of social policies' takeoff as the main explanatory factor that permits grouping the countries in the region. Accordingly, he categorizes them as pioneers, intermediate countries, or delayed countries. The timing is taken as a path dependence assumption for differences in benefit institutionalization and population coverage. His most important contribution was to outline stratification as the common mark of the Latin American model of social protection, being present in all the three types. As a consequence, even the oldest and more inclusive systems still keep different conditions of access and benefits, according to the bargaining power of each group covered.

Filgueira and Filgueira (2002) compared several countries, combining the amount of public social expenditure with the results expressed in social data, and reaffirmed Mesa-Lagos's conclusion on the Latin American pattern of social benefits stratification. In addition, they stressed the combination of this stratified inclusion of formal workers with different degrees of social exclusion. This result cannot be explained solely by the economic development of the country or through the maturation of the systems. Instead, it seems to be mostly dependent on political options. They found, until the 1970s, three different groups of welfare benefits distribution: a group of stratified universalism (Argentina, Chile, and Uruguay with high public expenditure and almost universal stratified coverage), a group of dual system (Brazil and Mexico, with high public expenditure and low coverage), and a big group of systemic exclusion (where most of the population was not encompassed). The Brazilian case is emblematic of how one of the pioneer countries in social protection can continue to present a high degree of social exclusion, in spite of the economic development or the social expenditure level, during the period considered in this study, from the beginning of the twentieth century up to the 1970s.

Analyzing the proliferation of social reforms in the region after the 1970s, Fleury (2001) observed that the fragile institutionalization of social rights and the explosion of massive urban demands were responsible for transforming the region into a social laboratory, where many different projects were designed and implemented. These reform efforts—in health care and in social security systems—have been part of the changing context brought about by the democratization of the region's political systems, the updating of its economic productive models, and the redesigning of the state's role. All of these have served as means to address the fiscal crisis and create the necessary conditions for positioning the regional economies more advantageously in the integrated and competitive global market.

In this context, there has been a change in the political structure around social protection policies with traditional powerful actors like bureaucrats and trade unions losing control over the stratified social security system. A new political arena was formed as a consequence of the introduction of market mechanisms in social policies as well as social movements' demands, generating a complex web of relations among financiers and providers, public and private actors.

As a result of this changing process, Fleury (2001) identified some paradigmatic reform models in the region, drawing special attention to the political coalition that supported the reform project, as well as the relationships among public and private agencies. The first type is the *market-oriented reform*, based on the Chilean experience. Its design generated a dual model of health care and pensions systems, with the segregation of the poor in the public sector and the affiliation of those who can afford it in a flourishing insurance market. The new social market was created by compulsory public policies, not a result of voluntary individual or corporative affiliations. Different from the traditional Bismarckian social insurance corporatist model, this structural reform inaugurated a typically market-oriented model, based in the purchasing power of the consumer, with the individualization of risks and the total absence of solidarity bonds among the beneficiaries.

The second type is the Colombian reform, based on the same market competition principle, although trying to reduce the inequalities of the Chilean experience by introducing some mechanisms of solidarity and state control of the per capita cost of benefits, in order to counterbalance the tendency to exclude some disadvantageous groups. They opted for a *competitive insurance system* with the participation of both private and public resources, thus forming a pluralistic model based on managed competition regulated by the state. This model places the insurance market at the core of the social policies' systems. A package of benefits is guaranteed either by the individual's own contribution or through the collective contribution to a solidarity fund. In the last case, the package of benefits is less generous than in the first, and the inclusion of the poor depends on the amount of resources available yearly. The complex network of public and private interests with diverging logics was unable to overcome the constraints of this design, which is based on limited salary contributions and on a savings-oriented logic inherent to the profit-seeking insurance companies. The influence of private interests had many undesirable effects on hospital and primary health care.

The third model was based on the Brazilian experience of creating a public universal system of integral social benefits in charge of assuring social rights through a more democratic, equitable, decentralized, and participatory system of social policies. This protectionist model was inspired by the social democratic experiences, although it innovated by means of a set of participatory mechanisms. Nevertheless, this public system was, from the beginning, highly dependent on private facilities because the dictatorship had already privatized the public health services. The Brazilian experience clearly denies the possibility of lumping the Latin American reforms in the same basket of liberal WS in the region, because it follows a social democratic model. Besides their differences, they faced similar challenges to build a universal WS, regarding the scarcity of resources for funding social systems and the existence of strong disparities among municipalities (Trydegarard and Thorslund 2010).

The three different social policy reform designs in Latin America can be named, respectively, as dual, plural, or universal (Fleury 2001), according to their proposals, contents, instruments, public-private relations, and supportive coalitions. A crucial variable to explain these different options is the timing of the reform with respect to two main macro processes: the economic crisis with the ensuing macrostructural adjustment, and the transition to democracy and the outburst of a new political fabric filled with social demands. It explains the prevalent values and principles as well as the material possibilities of each paradigmatic reform.

BRAZIL'S TRAJECTORY OF SOCIAL PROTECTION

All Latin American societies are assuming a new profile, with a more pluralistic and comprehensive system of social protection, with different arrangements of public and private facilities and functions. As a result, although the coverage is increasing in the region, there are many different conditions to access benefits. Whereas the benefits were determined by the purchasing power of each group, the result is to enhance social segmentation, even with the emphasis on public policies to target the poor. Instead, the universal social rights system could be an instrument to build a more cohesive society and expand citizenship status in order to overcome the traditional stratification and social exclusion. The condition here is the guarantee of access and quality for all.

The singularity of the Brazilian case, in which the design of a universal system took the opposite direction of the general trend of liberalism, can be explained only based on the close connection established between this project and the democratization process, in a context of high social mobilization and low perception of the economic constraints. In the mid-1970s the struggle for the democratization of policies takes on new characteristics and strategies. Whereas before it was confined to universities, clandestine parties, and social movements, from then on it begins to be located at the center of the state itself, with a strong push towards reforms. Rescuing the so-called social debt of redistribution and recognition became the central theme of the democratic agenda, gathering social movements of diverse natures. This process was intensified in the 1980s with the rise of a rich social fabric based on the union of the new syndicalism and the social rural and urban movements (mostly supported by the progressive wing of the Catholic Church), as well as on the emergence of leftist parties. In the fight for the BWS, one of the most distinguished actors was the Sanitary Movement, a coalition of political actors that was capable of formulating a consistent project for the new health care system. This same design was further applied to all social policies.

The new social protection design was rooted in a strategy to struggle against the dictatorship regime by democratic means, adopted by the leftist social movements in Brazil, mostly influenced by the Communist Party— which had strong presence among the intellectual elite up to the 1970s— and Catholic popular organizations.

Beyond that, the leftist movement was influenced by the concept of hegemony in Gramsci's works, and the goal of achieving hegemony in a complex civil society oriented the political movement to organize the society around some comprehensive flags. Further, Poulantzas's ideas on the importance of the political fight inside the state apparatus also influenced the movement and refreshed the Marxist conception of class struggle, leading the leftist groups to give priority to achieving strategic positions, in a piecemeal reforming process.

Differently from other left-wing parties and movements in Latin America, it must be recognized that the Brazilian opposition opted to take the risks inherent to this reformist strategy. By assuming important positions in governments of wide political coalitions, the leftist movements endorsed participation in the reform process despite the risk of either being co-opted by the political elites or legitimating an unsuccessful transformation.

During the last two decades, Brazil has been building a WS based on the principle of extending social rights to all citizens through universal social policies—a principle that is enshrined in the social security concept of the Federal Constitution of 1988. The struggle for a new comprehensive social protection model had an original component of social mobilization in favor of expanding social rights as part of the transition to democracy. The singularity of having a social policy project designed by social movements and its strong association with the transformation of the state and society into a democracy added some important characteristics to the BWS. The outstanding features are the combination of a highly decentralized system with a decision-making process that incorporates the federative system and organized society in participatory arenas. The existence of political spheres for negotiation and instruments for consensus building were important innovations in the intergovernmental relationships and in the relationship between the state and social actors.

Given the fragility of the political party system, social movements assumed the task of projecting new policies and public structures, under the flag of universal citizens' rights. In a society that is marked by profound inequality, the defense of equalitarian policies represented a counterhegemonic banner for the democratic project.

Given the previous experience of expressive economic growth, which did not lead to redistribution, it became clear that state-driven economic development was insufficient to assure public interests. Thus civil society organizations prioritized the construction of a democratic and inclusive nation in their commitment to transform existing public policies. All this democratic effervescence was channeled to the National Constituent Assembly, whose works began in 1987. The Federal Constitution of 1988 represented a profound transformation of Brazil's social protection model, legally consolidating the social rights in response to the pressures that had already been felt over a decade. A new period was inaugurated, in which the model of social security started to structure the organization and format of BWS towards a universal citizenship.

The Constitution of 1988 advanced in comparison to the previous legal formulations by guaranteeing a set of social rights, expressed in the innovative Social Order Chapter, which includes education, environment, culture, sports, minorities, communication, science and technology, and the social security structure. This social security arrangement was recognized as the core of the BWS. It was defined as "an integrated set of initiatives by the public power and the society, destined to ensure rights related to health, social insurance and social assistance" (Title VIII, Chapter II, Section I, art. 194). Therefore, the inclusion of social insurance, health care, and social assistance as parts of social security introduces the notion of universal social rights as an integral part of citizenship.

This new arrangement innovated in separating, for the first time, the social order from the economic one, which gives to the former the same status and priorities usually given to the economic field. The primacy given to social rights became evident with this autonomy and with the creation of specific contributions for funding social policies. As a block grant they constitute a unique social security budget, created apart from the fiscal national budget. These resources would be managed by the Social Security Council, integrated by the three ministries and representatives of the users.

The new constitutional social policy model is characterized by the *universality* of coverage, the recognition of *social rights*, under the guarantee and *duty of the state*, and the subordination of the private sector to regulation based on the *public relevance of actions and services* in these areas. The new public arrangement is a *decentralized public* network cooperatively *managed by participatory mechanisms*.

The originality of the Brazilian social security lies in its strong state reform component, in redrawing the relationships between the three levels of government and creating arenas for consensus building. The reshaped federalism addresses the main responsibility in the delivery of social policies to the local authority. The organization of the social protection systems should adopt the format of a decentralized, integrated network, with a single political command and a unique fund in each sphere of government, regionalized in a hierarchical arrangement. Moreover, it also created deliberative arenas, with equal participation of civil society and government representatives in each sphere of administration.

Although the stimulus for decentralizing was top-down, it provided a common guideline to public policy and to strength local governance by increasing administrative skills, as a condition for the acquisition of more resources and management autonomy. Furthermore, the design of the decentralization process included original elements, namely the existence of a participatory mechanism in each administrative level. In this sense it required a twofold movement, from the central administrative level to the local levels and from the state to the society. The intersection of these two lines gives place to a new model of local governance, with the existence of important mechanisms of consensus building, social control, and policy formation.

The main mechanism for decision making and consensus building among the three levels of government with concurrent competences concerning social policies is the Tri-Partite or Bi-Partite Commissions, encompassing, respectively, three or two governmental levels. This innovation has been considered a major advance in the country's federalist design, because it was an effective channel to deal with the inherent conflicts during the negotiation of policies, as well as the establishment of norms and parameters for resource allocation.

There are two participatory mechanisms that include both governmental authorities and the population: Councils and Conferences. The Councils exist in each level of the system and are mechanisms of social control, defining policy priorities, in charge of surveillance and budget approval, and evaluating executive proposals and performance. The Conferences, on the other hand, are called periodically to discuss agendas and convey different interests to a common platform, and in this way represent the main mechanism to form public values. Many experts discuss whether the Councils effectively have capacity to control the government, and the conclusion is less clear-cut than one might desire, due to the great diversity inside the country and the unequal distribution of resources among the participants. But studies proved that participation in the Councils is a resource to insert the councilor in important power networks and also an arena to challenge the control of organized groups, like political parties and union members (Côrtes 2009).

In spite of the limitations discussed above, the participatory mechanisms described have without doubt made an important democratic contribution and promoted the enlargement of the public sphere, restricting the traditional elite. The construction of dialogic mechanisms generates the recognition of popular actors and demands, and might permit a renovation of the political elites, with the incorporation of new members. The development of the WS was related to the changes from a liberal capitalism to a neo-corporatist form of relations among collective actors under state regulation. The differentiating factor of late-developing societies like Brazil is the composition of the civil society, where few trade unions share the participatory space with social movements and other civic organizations. Besides the heterogeneous composition of civil society, another peculiarity is related to the introduction of mechanisms of participatory democracy. Whereas in a traditional political conception participation is limited to the electoral process and redistributive policies, in the Brazilian case a great emphasis is given to the construction of comanagement instances, where state and society contribute to the policy-making process and implement mechanisms of social control. Although this still falls short of a deliberative democracy in which the collectively made decisions are mandatory upon the government, there is nevertheless an empowering process in progress as part of this statecraft blueprint.

FROM RHETORIC TO REALITY

During the 1990s Brazil experienced a hard period of economic crisis with hyperinflation and a great amount of foreign and internal debt accumulation, having to launch many different policies and instruments to cope with this situation that started to succeed only after the middle of the decade. Pressures from IMF to cut public expenditure, privatize public services, and downsize the state were strongly felt. At the same time, the government had to deal with social pressure to implement the transformations requested in the constitutional frame. In other words, there were two concurrent movements going in opposite directions: one expressed by the macroeconomic adjustment measures and the other by demands for assuring social rights and institutionalizing the WS. The following analysis will focus on this contradictory process of building a comprehensive and universal system under such adverse economic conditions, taking into consideration other adverse cultural and organizational factors that reduce the possibilities to achieve the equalitarian principle of citizenship.

Economic Constraints

One of the basic assumptions underlying the development of the WS in European countries is the synergetic influence of economic growth and social inclusion achieved through redistributive policies. The Keynesian model implies a strong presence of public policies in order to balance supply and demand in an affluent economic cycle. In the absence of economic growth to handle the bulk of the social demands, Brazil had to develop an institutional framework for social protection during the crisis of the 1990s. In this decade, the economic situation of the country moved from a turbulent scenario of high inflation and elevated debt to another one where the measures adopted to control inflation, such as the maintenance of mammoth interest rates continued increasing the fiscal debt. Without sustainable economic growth and the inability to count on public investments, unemployment in metropolitan areas doubled during 1995–2003, and informality reached more than 45 percent of the working force (IPEA 2005).

The consequences of this poor economic performance were immediately felt in the social insurance coverage, which decreased from 66.6 percent of the working force in 1992 to 61.7 percent in 2002, and increased again from this year onward, reaching 65.9 percent in 2008 (Schwarzer 2009 :75). Only in the five last years (2005–10) did the labor market grow sufficiently to achieve the lowest unemployment rate in many decades, about 6 percent per year.

From the literature on WS, one can draw another requirement for building a redistributive policy—the existence of a fair and progressive tax system. So far, in the Brazilian case, all governments failed to promote a major tax reform, only revamping the old system, which overburdened the poor. The result is that whereas the poorest parcels of the population—encompassing families that earn up to two minimum wages—spend 48.8 percent of their revenue with taxes, those families that receive more than 30 minimum wages spend only 26.3 percent with taxes (IPEA 2009). In 2008 the total burden of taxation rose to 36.2 percent of the GDP and continues to expand, mostly due to the increase in taxation upon consumption, representing 59 percent of the total collected in 2007 (Salvador 2008), instead of mostly on revenue and wealth. This problem is pointed out by an OECD publication (2011) affirming that after the tax and transferences, the European inequalities are reduced by 19 points in the Gini index, whereas in Latin America they are reduced by only two points, because the tax system is highly regressive.

The government's possibilities to invest in infrastructure and social services were restricted by the mandatory production of primary surplus in order to balance a currency jeopardized by the huge public internal debt. Between 2003 and 2010 the public debt increased two and a half times, in spite of the effort to convey funds to the payment of the debt services. For the national budget for 2010, whereas 44.93 percent of the resources were assigned to pay the debt services and 22.12 percent for pensions, only 3.91 percent were consigned for health and 2.74 percent for social assistance (http://www.auditoria-cidada.org.br, February 2011).

Political Priorities

These economic constraints strongly affect the implementation of the WS in terms of its capacity of funding and financing. As a result, the deterioration of public services and professional carriers impedes universal access with quality. The coverage expansion through the universal systems of health care and primary education was not followed by the necessary resources, as expressed by the public social expenditure, impeding either the expansion of the network facilities or the improvement in the quality of the services.

Although from 1995 to 2005 the federal social expenditure rose 74 percent, the increase is unequally distributed among the different social policy areas. Whereas social insurance expenditure grew from 44 percent to 51 percent and social assistance had an increase of 1 percent to 6 percent, federal health expenditure diminished from 16 percent to 11 percent and education expenditure was reduced from 8 percent to 6 percent (Castro 2009: 111). This data reveals that the universal policies of health care and basic education have not been prioritized by governments in recent years. Poverty reduction and social insurance were the main priorities. On one hand, poverty alleviation was adopted as the foremost priority of the governments, what led to a redistribution of public resources. On the other hand, the contractual character of social insurance implied an increased volume of expenditures with pensions. Because most of these benefits are linked to previous contributions, this social policy tends to reproduce market inequalities. To provide poor people with better qualifications, other important policies were launched, such as quotas and special credit for poor students to enter into the universities, as well as measures to formalize microbusiness and to assure microcredit to the poor. They were responsible for expanding the national market and certainly will have an impact on the development of the new middle class.

Whereas the social security (SS) paradigm of universal social rights was supported by social movements, it was not considered as priority by the democratic governments. Financial restriction combined with the absence of governmental priority undermined the constitutional precept that defined SS as an integrated set of initiatives concerning the assurance of social rights related to health, social insurance, and social assistance. To guarantee this arrangement the Constitution mandates the creation of the SS Budget managed by the SS Council. From the beginning, these two mechanisms of planning and integrating policies did not work properly. The budget became only a formal accounting instrument instead of a common integrated plan of resource allocation. Moreover, the government introduced a mechanism to retain 20 percent of the SS resources to assure the payment of the public debt. The National SS Council was dissolved in 1999, after some years of ineffective coordination and integration of the three sectors.

The SS legal framework was unable to cope with the different institutional trajectory of the three areas, clearly path-dependent on their previous history, formal legacy, and clientele. According to Sposati (2009: 176), public policies for pensions, health, and assistance differ in many aspects, including the following criteria: inclusion; the nature of the rights—labor, social, and human; market and philanthropy participation; public management experience; and the previous degree of institutionalization. One can also add the important difference of social coalitions formed around each political arena with differential capacities to veto or to launch new policies.

Rights without Benefits

The institutional fragmentation inside SS had harmful consequences. The abandonment of the integration of the three policy branches compromises other constitutional principles, such as integral care and universalism. An example of how social policies were left unfulfilled was revealed following the governmental efforts to transfer cash to target people who then used this benefit to acquire goods and services that should be part of their social rights, such as medicine (Lobato 2007).

The existence of distributive conflicts among the SS sectors and the relative incapacity of coordinating policies had a strong impact on the statecraft of the Unified Health System (*Sistema Único de Saúde* [SUS]). The health care facilities originally exclusively reserved for the social insurance clientele were merged with the public services from the Health Ministry in order to create the unified and universal health care system. Nevertheless, the resources to fund them did not come along. As a consequence, the Health Ministry went bankrupt in 1993 (Santos 2008) and could not invest to improve and amplify the services' network. A tax on financial transactions (CPMF [Contribuição Provisória sobre Movimentações Financeiras]) was implemented to fund the health sector. However, because it was not earmarked, only one-third of the resources were assigned to health whereas the government used the bulk for other expenses, until this tax was extinguished in 2009.

Since the year 2000 a new constitutional amendment (EC. 29) defines the responsibilities of each level of government to fund the health care system, but this law was only recently specifically regulated, in December 2011. During this decade the relative participation of national government in funding the health system decreased, whereas states' governments disguised their mandatory contribution within other expenses, not proper to health care. As a consequence, the financial burden was shifted to the local authorities in charge of health care provision. The main evidence of this financial restriction is that although the Brazilian expenses with health care represent more than 8.5 percent of the GNP, almost 60 percent of this amount is private expense, which is not compatible with the international parameters for a universal public system.

The proposal of a public national health system was the most radical *decommodification* project in the design of democratic Brazilian social policies. With a public expense of only 3.5 percent of GNP, the public health sector is nevertheless responsible for the preventive care and surveillance for all inhabitants and delivers services for over 70 percent of the population (reaching 90 percent in the poorest regions). The whole population was given the right to the public health sector, despite the extremely difficult financial situation. However, problems of access and low quality in public services reinforce the existing inequalities between elites and poor people, in terms of capacities and status, undermining the goal of universal citizenship.

This situation has led the middle class to escape from the public system by contracting private insurances mostly through corporation benefits. Although the coverage of the business insurance does not go beyond 30 percent of the population, pressure groups, like trade unions, push towards this market solution, and new popular plans were recently launched to attend to the demand of the lower middle class. The national government itself provides access to private insurance for civil servants and subsidizes the middle class by considering expenses with private pensions and health insurances as tax deductible. All these policies strengthen the private health care market and contribute to debilitating the public sector. Moreover, the public sector also subsidizes the private insurance companies because their beneficiaries have access in the public services to complex procedures not covered by their plans. By contrast, poor patients have to commute from one public service facility to the other in order to have access, thus revealing the gap between rights and benefits.

Benefits without Rights

There is no doubt that the most impressive impact of the social policies' designs after the Constitution of 1988 was integrating social assistance into citizenship status. Similar to the health system, an institutional system for social assistance policies was created, based on the same principles of assuring social rights through a public national network with different and articulated levels of government. This turning point in the social assistance policy was not followed by a straightforward institutional process. Initially, there was a clear governmental refusal to implement this constitutional mandate and to create the SUAS (Sistema Único de Assistencia Social [Unified Social

Assistance System]) in the same participatory and decentralized form as the health system. Only after strong pressure from civil society and professional organizations did the government pass the social assistance law in 1993 (LOAS [Lei Orgânica da Assistência Social]), the first step in the institutionalization process. But from 2000 onward the public agenda concerning social assistance has diverged from guaranteed citizen rights within the Constitution to a targeted model to fight poverty. From then on, the country had two branches of social assistance policies, constitutional and targeted.

The constitutional model is characterized by the creation of a special regime for noncontributive pensions encompassing benefits for all rural workers, the urban impoverished elderly, and handicapped people. In this last case pensions depend on a means test and come from the Continuous Cash Benefits (BPC [Beneficio de Prestação de Continuada]) program. The targeted model's main feature is the Family Grant Program (Bolsa Família [BF]), which offers conditional cash transfer to the poorest families. For Lobato et al (2009: 722) the permanence of these hybrids jeopardizes the fairness of the citizenship model proposed in the 1988 Constitution. Whereas the Continuous Cash Benefits program is a legal right, the Family Grant is a discretionary benefit and the conditions of access and the value of the benefits vary greatly. Although the access to the Continuous Cash Benefits program is much more restricted than in the Family Grant,[1] the pensions legally guaranteed by the Continuous Cash Benefits program are linked to the national minimum wage and are three times the average value of the Family Grant benefit. On the other hand, whereas BPC covers 3.8 million people, Family Grant reaches over 12 million, although the Census of 2010 pointed out that one out of four rural residents is still living in indigence. The existence of conditional requirements for cash transfer in the Family Grant Program, besides the means proof, as well as the absence of a legal status for the beneficiaries widened the gap between the two main welfare benefits. Although the management of the program is based on technical criteria, the absence of formal rights to welfare benefits is considered as an open space for clientelism (Wood and Gough 2008: 321).

Brazil has a high coverage level for elderly people (81.7 percent), thanks to the efforts to include them in noncontributive benefits. Because there was a sustained policy of increasing the minimum wage in the last decade, significant income redistribution resulted. Nonetheless, the existence of about 34.1 percent of workers not covered by social insurance represents an important challenge to the objective of a universal WS (Schwarzer 2009: 73).

Nonetheless, the impact of social protection on poverty reduction is one of the most outstanding results of social policies. Associated with the economic growth and the recovery of the minimum wage's purchase power, social policies are restratifying the Brazilian society. These recent phenomena have provoked the enlargement of the middle class with the upward mobility of millions of families from the lowest stratum to the middle class (Neri 2010).

The synergetic combination of distributive policies and economic growth has a stupendous effect on income distribution, and the consolidation of a middle class is not only a product of democracy but a central requirement for its sustainability. The expansion of the national market through the enlargement of the middle class was one of the powerful assets that allowed Brazil to cope with the economic crisis. Nevertheless, guarantying good quality of services in the education and health sectors for the emergent class into the universal systems of social protection is the new challenge. Institutionalizing democracy and expanding possibilities to a sustainable economic growth depends on the capacity to build new skills and to assure citizens' rights, reducing inequalities.

CONCLUDING REMARKS

Based on the Brazilian experience, Fleury (2009) argued that different components of the democratic consolidation process have distinct paces despite being part of the same movement. This may lead to synergies and complementarities, but also tensions. The analysis of the reforming process was based on the identification of three inherent components and their dynamics: the process of *subjectivation*, through the construction of political subjects, social actors, and their political coalitions; the process of *constitutionalization*, through which social rights gained their legal framework; and the *institutionalization* of social policies via implementation of the correspondent structures and mechanisms for accomplishing the functions of financing, provision, regulation, and system modulation.

The context were the processes of subjectivation and constitutionalization that occurred that differed from the economic constraints that occurred during the institutionalization of the WS, justifying the inobservance of constitutional prescriptions specifically related to the resources assigned to social security policies. Beyond that, the legacy of a former corporatist social protection model enabled the maintenance of social insurance as the dominant policy, reserving the bulk of the funds for the less egalitarian policy. The different legacies of the three social security branches in terms of knowledge and capacities, degree of institutionalization, political supporters, governmental priorities, and relations with the private sector were influencing factors of the path followed by them. Another important issue is the political presence and strength of social movements in the defense of social rights.

The advances in the institutionalization of the constitutional model could be observed in several aspects of the three policies, as the introduction of noncontributive pensions for rural workers, cash-transferring programs to fight poverty, and the expansion of the health system in small municipalities. The most meaningful transformations were in the state structure, with the introduction of new instruments of the federalist organization associated with mechanisms of social control and participation.

The restrictions placed on the full institutionalization of the constitutional guidelines open up space for the progressive introduction of other interests and structures in the original design, reshaping the WS proposal in practical terms. This did not result in a pure productivist model, because there was not a structural change in the constitutional model that afforded universal rights. However, governmental priorities gave rise to a hybrid design in social assistance, in which cash transfer programs and legal benefits coexist with differences in access and rights. Regarding the health sector, the strength of the movement in favor of the universal system was strong enough to avoid pressures to privatize it, as occurred in many countries of the region, but was unable to avoid the surfacing of hidden mechanisms that progressively introduced new flows of resources, patients, and doctors between the public and the private sectors. Instead of a rupture with the constitutional SS design, changes were implemented via piecemeal transformation giving rise to a mixed system combining universal social rights with a flourishing market of privately provided health goods and services, in which the public and the private are linked through hidden mechanisms and association.

As a result, we can find dual anomalies in the BWS: rights without benefits (in case of nonaccess) and benefits without rights (in conditioned cash transfer). This occurs despite the positive outcomes from the socially targeted policies such as the current trend of upward mobility. This phenomenon is also the consequence of the recent period of economic growth and the priority given to the fight against poverty in the governmental agenda. However, the growing inclusion of the poorest populations in the market is not enough to return to the constitutional model. To do so would require promoting their inclusion in the public sphere by assuring them full citizenship status and the skills to insert them in the labor market under better conditions. Without the social and political inclusion through social rights and redistributive policies that are capable of assuring better living conditions for the population, it will be impossible to reduce urban violence and other symptoms like low cohesion in Brazilian society.

The quality of the public universal social protections system of health care and education will be the next challenge for BWS consolidation. But it will require a big change in the governmental priorities, assigning financial resources and implementing performance-oriented management in order to achieve better-quality standards and effective access in social services. Whereas the government insists in considering social policies narrowly as social expenses, this change is far from being implemented. Social investments are essential to improve productivity, develop new technologies, and reduce the national dependence on monopolized foreign imports. Beyond this, they mobilize a mass of skilled workers, with great diffusion and leadership in their communities. Last, but not least, it is necessary to change the dominant elitist culture, because it tends to perpetuate injustices and discriminations as common forms of treatment of minorities and poor people, thus reproducing differential access and quality of public goods consumption. These micro politics act as a counter-right in practical terms. The fight against social injustices requires a permanent process of political mobilization to face the challenge of transforming the rights-in-principle into rights-in-practice (Foweraker and Landman 1997:19).

NOTES

1. Eligibility: Twenty-five percent of the minimum wage per capita income for BPC and less than two American dollars daily per capita income for BF. While the BPC has an automatic adjustment BF adjustment is not automatic. In 2013, BPC benefit value is near tree hundred fifty dollars while BF basic plus variable benefits can go from eighteen dollars to the top of one hundred eighty dollars.

REFERENCES

Barrientos, Armando. 2008. "Latin America: Towards a Liberal-Informal Welfare Regime." In *Insecurity and Welfare Regimes in Asia, Africa and Latin America: Social Policy in Development Countries*, edited by Ian Gough, Geof Wood, Armando Barrientos, Philippa Bevan, Peter Davis, and Graham Room, 121–68. Cambridge: Cambridge University Press.

Castel, Robert. 1995. *Les Métamorphses de la Question Sociale*. Paris: Gallimard.

Castro, Jorge. 2009. "Política Social: Alguns Aspectos Relevantes para Discussão." In *Concepção e Gestão da Proteção não Contributiva no Brasil*, by Brasil, MDSCF (Ministério do Desenvolvimento Social e Combate à Fome), 87–132. Brasília: UNESCO.

Côrtes, Soraya. 2009. "Conselhos e Conferências de Saúde: Papel Institucional e Mudança nas Relações entre Estado e Sociedade." In *Participação, Democracia e Saúde*, edited by Sonia Fleury and Lenaura Lobato, 102–28. Rio de Janeiro: Cebes.

Esping-Anderson, Gosta. 1990. *The Three Worlds of Welfare Capitalism*. Cambridge: Polity Press.

———. 1995. "O Futuro do Welfare State na Nova Ordem Mundial." *Lua Nova: Revista de Cultura e Política* 35: 73–112.

Filgueira, Carlos, and Fernando Filgueira. 2002. "Notas sobre Política Social em América." Mimeo. Washington, DC: INDES, BID.

Fleury, Sonia. 2001. "Dual, Universal or Plural?: Health Care Models and Issues in Latin America: Chile, Brazil and Colombia." In *Health Services in Latin America and Asia*, edited by Carlos Molina and Jose Nuñez del Arco, 3–36. Washington, DC: Inter-American Development Bank.

———. 2009. "Brazilian Sanitary Reform: Dilemmas between the Instituting and the Institutionalized." *Ciência & Saúde Coletiva* 14(3): 743–52.

Flora, Peter, and Jens Alber. 1981. "Modernization, Democratization, and the Development of Welfare States in Western Europe." In *The Development of the Welfare States in Europe and America*, edited by Peter Flora and Arnold Heidenheimer, 37–65. New Brunswick, NJ: Transaction Press.

Foweraker, Joe, and Todd Landman. 1997. *Citizenship Rights and Social Movements: A Comparative Statistical Analysis*. Oxford: Oxford University Press.

IPEA. 2005. *Radar Social*. Brasília: IPEA –instituto de Pesquisa Econômica Aplicada.

———. 2009. *Receita Pública: Quem paga e como se gasta no Brasil*. Comunicado da Presidência IPEA—Instituto de Pesquisa Econômica Aplicada, Brazil, June 30.

Kim Sung-Won. 2009. "Social Changes and Welfare Reform in South Korea: In the Context of the Late-Coming Welfare State." *International Journal of Japanese*

Sociology 18 (1): 16–32. http://onlinelibrary.wiley.com/doi/10.1111/j.1475–6781.2009.01116.x/pdf (accessed Jan/12/2011).
Lobato, Lenaura. 2009. "Dilemmas of the Institutionalization of Social Policies in Twenty Years of the Brazilian Constitution of 1988." *Ciência & Saúde Coletiva* 14 (3): 721–30.
Lobato, Lenaura, et al. 2007. "Avaliação do Benefício de Prestação Continuada." In *Avaliação de Políticas e Programas do MDS—Resultados*, Vol. 2, organized by Jeni Vaitsman and Romulo Paes e Souza, 257–84. Brasília: MDS/SAGI.
Marshall, Thomas H. 1967. *Cidadania, Classe Social e Status*. Rio de Janeiro: Zahar Editores.
Mesa-Lago, Carmelo. 1978. *Social Security in Latin America: Pressure Groups, Stratification, and Inequality*. Pittsburgh, PA: University of Pittsburgh.
Neri, Marcelo Côrtes, coord. 2010. *The New Middle Class in Brazil: The Bright Side of the Poor*. Rio de Janeiro: FGV/IBRE, CPS.
OECD. 2011. "Growing Income Inequality in OECD Countries: What Drives It and How Can Policy Tackle It?" OECD Forum on Tackling Inequality, Paris, May 2, 2011.
Offe, Claus. 1984. "A Democracia Partidária Competitiva e o Welfare State Keynesiano: Fatores de Estabilidade e Desorganização." In *Problemas Estruturais do Estado Capitalista*, 356–85. Rio de Janeiro: Biblioteca Tempo Universitário.
Oliveira, Francisco. 1988. "O Surgimento do Antivalor." *Novos Estudos CEBRAP* 22: 9–28.
Rimlinger, G. 1971. *Welfare Policy and Industrialization in Europe, America and Russia*. New York: Wiley.
Salvador, Evilásio. 2008. "Questão Tributária e Seguridade Social." In *Debates Contemporâneos Previdência Social e do Trabalho*, edited by Eduardo Fagnani, Wilmês Henrique, and Clemente Ganz Lucio, 387–402. São Paulo: Unicamp, CESIT, LTR Editorial.
Santos, Nelson R. 2008. "Vinte Anos do SUS: Por Onde Manterás Chamas da Utopia?" In *Os Cidadãos na Carta Cidadã*, Vol. 5, edited by Bruno Dantas et al., 146–68. Brasília: Senado Federal.
Schwarzer, Helmult. 2009. *Previdência Social: Reflexões e Desafios*. Brasília: Ministério da Previdência Social.
Sposati, Aldaíza. 2009. "Seguridade Social e Inclusão: Bases Institucionais e Financeiras da Assistência Social no Brasil." In *Seguridade Social, Cidadania e Saúde*, organized by Lenaura Lobao and Sonia Fleury, 173–88. Rio de Janeiro: Cebes.
Takegawa, Shogo. 2009. "International Circumstances as Factors in Building a Welfare State: Welfare Regimes in Europe, Japan and Korea." In *International Journal of Japanese Sociology* 18: 79, 96. http://onlinelibrary.wiley.com/doi/10.1111/j.1475–6781.2009.01119.x/pdf accessed Jan/12/2011).
Titmuss, Richard. 1958. *Essays on the Welfare State*. London: George Allen and Unwin.
Trydegard, Gun-Britt, and Mats Thoslund. 2010. "One Uniform Welfare State or a Multitude of Welfare Municipalities?: The Evolution of Local Variation in Swedish Elder Care." *Social Policy and Administration* 44 (4): 495–511.
Wilensky, Harold. 1975. *The Welfare State and Equality: Structural and Ideological Roots of Public Expenditures*. Berkeley: University of California Press.
Wincott, Daniel. 2011. "Images of Welfare in Law and Society: The British Welfare State in Comparative Perspective." *Journal of Law and Society* 38 (3): 343–75.
Wood, Geof, and Ian Gough. 2008. "Conclusion: Rethinking Social Policy in Development Contexts." In *Insecurity and Welfare Regimes in Asia, Africa and Latin America: Social Policy in Development Countries*, edited by Ian Gough, Geof Wood, Armando Barrientos, Philippa Bevan, Peter Davis, and Graham Room, 313–26. Cambridge: Cambridge University Press.

2 Inequality, Poverty, and the Brazilian Social Protection System

Marcelo Medeiros, Sergei Soares, Pedro Souza, and Rafael Osorio

At first glance, Brazil seems to be resolutely swimming against the tide when it comes to social protection: the 1988 Federal Constitution—enacted during a period of reform and liberalization elsewhere—had a very clear social democratic slant, enshrining social rights typical of Western European welfare states and beginning a break with a strongly corporatist heritage. It established a three-pronged social security network, composed of a free and universal health care system (the Unified Health System, or SUS), an expanded pensions system responsible for retirement pensions and other work-related insurances, and a social assistance system meant to provide services and cash transfers to very low-income families.

A more detailed view, however, shows a far more convoluted picture. Although the Constitution was certainly a turning point, social security policies are still an evolving patchwork, riddled with inequalities. As usual, demands for rationalization and redistribution are often somewhat counterweighted by vested interests, so that, from the point of view of poverty and inequality reduction, the overall system tends to improve at a glacial pace. Policy makers usually have to accommodate well-meaning innovations with sizable concessions, which goes a long way in explaining the increase of government transfers from 11.7 to 14.2 percent of GDP between 1997 and 2008.

In this chapter we shall take a closer look at the pensions and social assistance systems and analyze their impacts on poverty and inequality. These two systems can be more directly related to the idea of a social protection system than any other, although such classification has no correspondence in the institutional, legal, or administrative organization of social policies in the country. In particular, we are interested in two aspects of social protection, the distributive effects of retirement pensions (for both the public and the private sector) and targeted cash transfers, and the impacts of the conditionalities of the Bolsa Família program, which is often taken as a reference for the design and implementation of targeted conditional cash transfers (TCCTs) programs elsewhere.

Given our focus, there are parts of the social protection system that will not be discussed, such as unemployment and severance benefits, and social assistance services. Likewise, we will not discuss the distributive effects of the taxes that fund this system. In any case, it is worth highlighting that retirement pensions and targeted cash transfers comprise the bulk of social protection expenditures.

THE SOCIAL PROTECTION SYSTEM

Social protection in Brazil is a combination of a core of national policies—pensions, unemployment insurance, and cash transfers—and a multitude of local actions loosely integrated. National policies are small in number but far more relevant in terms of budget and distributive impacts. This occurs because Brazil is a federation in which the federal government is much more powerful than either states or municipalities. On the one hand, the Constitution and other laws require that municipal, state, and federal governments share responsibilities, but on the other, the federal government retains half of total tax revenue and the prerogative to regulate and oversee policies and programs in almost all areas. We can thus conceive the existence of a unitary Brazilian social protection system if the objective is to examine the large-scale impacts of that protection on the well-being of the population, as we intend to do here, but this same idea would have to be taken with a grain of salt in the case of policies oriented to local problems or to particular social groups.

This system exists in its current state since the enactment of the 1988 Constitution, but its roots lie further in the past. Although isolated policies can be traced back to colonial times, social protection began in earnest following the beginning of industrialization in the 1920s. As in many other countries, state pension schemes were the first step, and social protection began as a synonym for a collective insurance against the risks related to the loss of the individual capacity to work of a household provider, be it for old age or other reasons for early retirement. These pension funds were initially organized by occupational categories, which were later aggregated according to sectors of the economy to which these categories belonged. With few exceptions, these funds covered only formal workers in the urban areas. Such organization both reflected and reinforced the strong corporatist bias that marked Brazilian social policies during the twentieth century. As any corporatist institution, the pensions funds tended to reproduce pre-existing inequalities.

The consolidation of a series of pension funds in a more organized system, however, occurred much later, in the 1960s. In a context of rapid modernization of the economy and urbanization of the labor force, the sectoral funds were unified into a national institution responsible for the administration of the pensions of most private sector workers, the Instituto Nacional de Previdencia Social (INPS). Some years later, social assistance acquired the status of a national policy with the inauguration of two programs: (1)

the today almost extinct Renda Mensal Vitalicia (RMV), a noncontributory monthly pension with a strong moralistic approach, designed to protect the 'deserving poor' among people with disabilities or aged seventy and older, and (2) the today still relevant Aposentadoria Rural, a noncontributory pension for rural workers established to protect those who remained in the agricultural sectors of the economy.

During the 1980s, Brazil left behind two decades of military dictatorship and this opened up the political environment for a more egalitarian agenda. As a result, a space for reform in the social protection system was created. The institutional landmark of these changes was the 1988 Constitution, and this new environment became clearly visible during the second half of the 1990s, when social democracy became a dominant political trend in the country. Different parties have alternated in power at national and subnational levels, but with remarkably similar policies due to the strong coalitions in society in their support, creating a favorable continuity for long-term reforms.

The 1988 Constitution introduced in the Brazilian law the concept of a social security system, putting health, pensions, and social assistance under similar guiding principles. The new Constitution not only set up general principles for social policies, it specifically determined how some policies should be designed. For instance, it established social assistance as a social right and created a minimum wage social assistance benefit for people with disabilities or the elderly living in low-income families. It also established the national minimum wage as a minimum limit for public pensions and social assistance. More than a symbolic move, this ensured that the power of labor unions and the popular appeal of increasing the minimum wages in a growing economy would become a political backup for social protection. And it was indeed: two decades later a considerable share of the persistent fall in inequality the country was due to the increases in the value of the pensions and assistance indexed to the minimum wage.

In addition to direct legislation on social protection, the new Constitution also established the legal basis for the institutions necessary to coordinate national, state, and municipal levels of government. On the one hand it decentralized the administration and the implementation of several social policies; on the other, it allowed the formation of the centralized funds that would maintain the policies in pace, irrespectively of the fiscal capacity of local governments, and also allow the federal government considerable leverage over states and municipalities. This centralized-decentralized arrangement became the backbone of many social policies. It was the coordination of national and subnational governments started in the mid-1990s that paved the way for the expansion of social assistance in the country. Sustained by ideological continuity over more than two decades, this coordination allowed the implementation of the large-scale antipoverty programs in the country, particularly Bolsa Escola and its rebranded version, Bolsa Família.

Currently, the state pension system itself is very heterogeneous. This occurs because, in spite of its egalitarian initiatives, the 1988 Constitution made corporatist concessions and did not unify the state pension systems for public and private sector workers. The lack of unification created two parallel pension subsystems or regimes in Brazilian terminology: the Regime Geral da Previdencia Social (RGPS), the general regime for the private sector; and the Regimes Próprios de Previdência Social (RPPS), a subsystem that provides pensions to public sector workers, including categories such as the military, teachers, health workers, and administration workers (legally speaking, it is a set of subsystems that operates under uniform rules).

Both subsystems are similar insofar as they are contributory and follow a pay-as-you-go scheme, mostly funded by labor contributions—with the burden shared by workers and employers. Additional funds used to keep the system under equilibrium come from taxes and respond for a minor part of the total budget of the RGPS and a much larger share of the RPPS.

However, there are stark differences between them. Private sector workers' pensions have a legal cap that limits their maximum value, which has important implications for income inequality in a country with an aging population. Public sector pensions have no such cap, in addition to other rules that are far more lax than for private sector workers; for instance, until the 1998 reform, civil servants contributed a much lower percentage of their earnings than private sector workers, they could accumulate benefits, they had pensions higher than their last full paycheck, and they often could retire earlier than their private sector counterparts.

Two presidents, Cardoso and Lula, manifested their intention of unifying the two subsystems but neither fully succeeded (Marques and Euzéby 2005; Melo and Anastasia 2005). Nevertheless, both implemented reforms that cut down many of the privileges of the RPPS. Only in 2012 has the full convergence of the two regimes begun, but, because new rules will apply only to new hirings in the public sector, full unification will not happen for decades.

Albeit more egalitarian, the RGPS of the private sector workers is also polarized. Almost two-thirds of benefits paid by RGPS are equivalent to one minimum wage. They usually are subsidized within the pay-as-you-go scheme and paid to people who were low-income workers, including former rural workers who benefit from the Aposentadoria Rural. These benefits are not as egalitarian as the social assistance transfers, but still are highly progressive. As a result of the large number of benefits distributed, these subsidized pensions are one of the most important anti-inequality policies Brazil currently has. Over the last decade the behavior of total inequality was strongly influenced by the variations on these pensions (Hoffmann 2003, 2007, 2009). The other benefits accrue to formal sector workers who had a better standing and thus receive higher pensions. As of 2012, the cap for the RGPS is R$3,916. This is slightly more than six minimum wages (R$622), but because the minimum wage keeps increasing in real terms and the benefit cap only keeps up with inflation, this ratio is steadily falling.

Despite recent efforts to expand the RGPS, participation in the system is still closely linked to formal jobs, which leaves a good fraction of the population unprotected. Autonomous workers have the option to contribute individually, but this implies payments from 11 to 20 percent of their income, a proportion high enough to prevent the participation of low-income workers. This design reproduces preexisting inequalities in two ways. Firstly, it does not offer protection to informal workers, who add up to about half of the labor force. Secondly, it pays benefits proportional to contributions, consequently perpetuating previous inequality.

Social assistance cash transfers are noncontributory and funded by taxes. They are all targeted to the low-income population, but the eligibility levels vary according to the program. There are two main targeted cash transfer programs in Brazil, the Continuous Cash Benefit (Beneficio de Prestacao Continuada [BPC]) and Programa Bolsa Família (PBF).

BPC transfers the value equivalent to the minimum wage to people legally considered incapable of working—persons with disabilities or aged sixty-five or more years—living in very low-income families. It is targeted and unconditional and its eligibility line is one-quarter of the going minimum wage. This means that it is doubly indexed to the minimum wage because both the value of the benefit and the number of beneficiaries are more or less linear functions of the going minimum. BPC is highly centralized, with states having virtually no role and municipalities helping only to identify beneficiaries in some cases.

Bolsa Família is a targeted and conditioned transfer composed of two benefits. Bolsa Família's variable benefit, currently (as of 2012) at R$32 per child (and R$38 per adolescent) in poor families, is at most a quarter of that of BPC, but its eligibility lines, currently at R$140, are higher. Bolsa Família's fixed benefit is R$68, given to extremely poor families. Because it is conditional, its beneficiaries must comply with some conditionalities, mainly that children attend school regularly and get their vaccines on time, and that pregnant women follow a schedule of prenatal exams. None of these conditionalities is seen as a burden to the families—some even argue they are unnecessary—because school attendance, vaccination, and prenatal examination levels in Brazil were already reasonably high before Bolsa Família was scaled up (Medeiros, Britto, and Soares 2007). States and municipalities are free to complement the federal benefit with benefits of their own, and most, although not all, such programs operate in coordination with the national program.

At the risk of oversimplifying, one can say that the value of the benefits and the eligibility lines make Bolsa Família an income *supplement* for families with capable workers, whereas BPC is an income *substitute* for individuals who cannot or should not have to work. Both programs are well targeted, highly progressive in distributive terms, and with a targeting efficiency similar to that of other Latin American cash transfer programs. Bolsa Família covers about one-third of the population of Brazil whereas

BPC covers ten times fewer people (about 3 percent of the population live in BPC families) (F. V. Soares, Ribas, and Osorio 2007; F. V. Soares 2011; S. Soares 2009; Soares, Ribas and Osorio, 2010; S. Soares et al. 2009).

IMPACTS ON POVERTY AND INEQUALITY

Overall Impact of Pensions on Inequality: RGPS and RPPS

Among the institutional determinants of inequality in Brazil, pensions deserve highlighting. From the legal point of view—and from the moral point of view too, some would argue —they should be egalitarian but, in practice, the pension system is a mix of redistributive and strongly regressive benefits whose final outcome is far from egalitarian. Second only to labor incomes in size, they are the most concentrated of all major sources of family income in Brazil. They contribute to about one-fifth of the total inequality in the country (Hoffmann 2003, 2009). An important exception is the subsidized pensions at the legal floor indexed to the minimum wage, which deserve further discussion below.

The rules organizing the public pension system make it replicate preexisting inequalities. Two articulated factors are behind this: pensions being proportional to contributions and the lack of a maximum limit for the value of benefits for former civil and military servants.

A contributory pension system calculates benefits taking into account the contribution made by workers over their active life. In a country that has been marked by extremely high levels of wage inequality over decades, this means that the power to contribute of workers was also very unequal. When the system pays pensions proportional to contribution it replicates previous inequalities. For instance, women who have been prevented from entering the labor market due to gender discrimination will be unprotected; those who received lower wages due to racial discrimination will also receive lower pensions as a result of that discrimination.

However, inside contributory pay-as-you-go pension systems, there are mechanisms to reduce this tendency to reproduce existing inequality. A minimum value or pension floor is one of them. By subsidizing pensions so no benefit remains below a certain level—the equivalent of the legal minimum wage, in the case of Brazil—the state reverts inequalities by paying more to those who had less contributory power. A maximum value or pension cap is another. Limiting the benefits tends to reduce the concentration of resources spent on public pensions on families who have other sources of income or who could have afforded private savings.

As a matter of fact, when these mechanisms are used, they do reduce inequality. Subsidized pensions at the floor of one minimum wage are highly progressive and became an important force driving inequality down in Brazil, as discussed below. Similarly, the cap on RGPS benefits ensures it does not become regressive. However, the RPPS of the public sector workers has no such limit and precisely because of this it is extremely regressive: RPPS benefits above the value that would be the limit of RGPS respond for much of its regressivity.

According to POF 2008–9, a consumption and expenditure survey, around 31 percent of the population of Brazil live in households where at least one person receives a pension. A considerable majority—28 percent of people in all households in Brazil and 90 percent of those in households with pension income—receives benefits from the private sector worker regime, RGPS. Only 4 percent of the population live in households benefiting from RPPS. However, the amount transferred from RPPS to this small group is almost half of the total expenditures of RGPS. Moreover, more than two-thirds of the money transferred by RPPS is received by less than 1 percent of population of the country, mostly families that would be considered wealthy even without the pensions. As a consequence of this dualism in the protection of workers from the public and private sectors, the share of total inequality related to benefits for those who were in the public sector is close to the share related to RGPS, 9 percent and 12 percent, respectively. RPPS incomes are so concentrated that the part of pensions above the limit imposed by RGPS, alone, contributes to 4 percent of total inequality as measured by the Gini coefficient, although less than 1 percent of the population receives it.

Impacts of Minimum Wage Pensions and BPC Assistance

The politics of determining the minimum wage have become very important for understanding the impacts of social protection on poverty and inequality and deserve a more detailed analysis. Due to so many social security and social assistance benefits being indexed to the minimum wage, it affects poverty and inequality not only through the labor market but also through the social protection policies.

The value of the national minimum wage is proposed by the president and approved by Congress. The real minimum wage has been increasing since price stabilization in 1994, and the rate of growth increased from 2005 onward. After negotiation with unions in 2007, a law passed early in 2001 rules that, up to 2015, the value of the minimum wage for the next year should be adjusted for inflation plus a real increase equal to the GDP growth two years before. Although diminishing returns have already began, this will cause the minimum wage to stay on as an important source of poverty and inequality reduction, at least in the near future.

As the value of the minimum wage rises, its role in social protection tends to increase over time. Because pensions just above the floor are adjusted annually for price changes only, but the minimum wage receives a real increase in addition to adjustment to inflation, more and more people will be receiving the pensions at the floor and this will reduce inequalities in pensions. As a matter of fact, this is already happening: the Gini coefficient of all RGPS benefits went from 0.37 in 2004 to 0.27 in 2012 (MPS 2004, 2012). In addition, as BPC sets a quarter of the minimum wage as its eligibility threshold, every time the real minimum wage is raised, the number of potential beneficiaries grows.

The contribution of minimum wage pensions and BPC to income inequality depends on their aggregate value and concentration. Because of the very high levels of inequality in Brazil, minimum wage pensions are concentrated on households in the middle of the income distribution, with few recipients in the bottom and on the top of it. As the real minimum wage increases more than other incomes, pensioners' households will advance their position in the income distribution. The incidence of BPC is predominantly among the extreme poor, but it channels the impact of changes in the minimum wage the same way: both programs are highly redistributive, although a bit less every time the minimum wage advances more than other income sources (S. Soares et al. 2010).

BPC and the pensions indexed to the minimum wage were very effective in shielding the Brazilian elderly from poverty. Whereas in 2004–9 the poverty headcount fell from 24 to 15 percent of the population, among those aged sixty-five or above it fell from 6 to 2 percent. Less than 0.5 percent of the elderly remained extremely poor in 2009 (Osorio, Souza et al. 2011). Given the current value of the minimum wage, households with elder members are, in practical terms, insured against extreme poverty. If the minimum wage continues to increase in real terms, as expected, soon they will be almost perfectly insured against (nonextreme) poverty as well.

Impacts of Bolsa Família

The impacts of Bolsa Família cash transfers on poverty and inequality are rather different from those of BPC, basically for two reasons. The first is its slightly more efficient targeting on the nonextreme poor, as its eligibility thresholds are higher than BPC's (S. Soares et al. 2009). The second is the low value transferred, insufficient to move many families out of poverty (Osorio, Soares, and Souza 2011).

As the households receiving it are poor and remain poor—or almost poor—the concentration of Bolsa Família is high on the bottom of the distribution. Indeed, it is the most redistributive of all major income sources in Brazil. Contrary to what happened with minimum wage pensions and BPC, the incidence of Bolsa Família at the bottom of the distribution increased from 2004 to 2009, whereas its weight on total income more than doubled. Therefore, it became even more redistributive, and in spite of its growth we still have not seen any diminishing returns. Although its weight is very low compared to other sources, less than 1 percent of total income, the incidence of Bolsa Família among the poor is so high that it became an important driver of the recent fall in income inequality in Brazil. Depending on the periods analyzed, around one-fifth of the fall is due to it (F. Soares et al. 2006).

Notwithstanding its impact on inequality, Bolsa Família is not very effective in eradicating poverty and extreme poverty. The lack of effectiveness is partly due to design: the value of the cash transfers is determined by the composition of a household. Although beneficiaries are divided into poor and extremely poor, within each stratum the value transferred is set regardless of their poverty gap. Two families with the same composition—one with income close to the extreme poverty line, the other without any income—receive the same benefit. Thus, Bolsa Família mainly alleviates poverty. Such a design reduces the poverty gap, but its impact on the headcount ratio is small: from 2005 on, the transfers reduced the extreme poverty headcount ratio by only 1 to 2 percentage points (S. Soares et al. 2010).

From 2004 to 2009, Bolsa Família cash transfers grew in importance as a source of income, particularly for the extremely poor households. As the minimum wage rose, most of the families that had a formal worker, or families that had an elder or disabled member receiving BPC, left extreme poverty. The remaining extremely poor families became almost entirely composed of informal workers, the unemployed, and inactive adults. The share of Bolsa Família transfers in the aggregate income of these families rose from 14 to 39 percent. At the same time, income sources of value greater than one minimum wage almost ceased to accrue to this stratum, falling to less than 1 percent of its aggregate income, and those equal to one minimum wage decreased from 17 to 6 percent (Osorio, Souza et al. 2011).

As Bolsa Família transfers more money when a family has people aged zero to seventeen years, its poverty reduction effect was stronger for families with children. Although children are still the age group most likely to suffer extreme poverty, the percentage of families without children grew from 16 to 25 percent among the extremely poor (Osorio, Souza et al. 2011).

EFFECTS OF CONDITIONALITIES

Whereas the previous sections of this chapter have focused on the impacts of all the transfers of the social protection system on income distribution, this section will concentrate exclusively on Bolsa Família's impacts on dimensions other than income inequality and poverty. Because it aims to be more than simply a cash transfer and change behaviors related to poverty, Bolsa Família have in fact drawn much more interest from many different audiences than those of retirement pensions and conventional assistance.

School Attendance Conditionalities

The school attendance conditionalities of Bolsa Família are of particular interest for the debate on long-term measures to reduce income poverty. The rationale for conditionalities comes from the thesis of an intergenerational cycle of poverty caused by a low demand for education in poor families.

The low demand in the present results in low education of the children of the poor and this will be a future cause of poverty. Bolsa Família intends to break this cycle by using conditionalities to induce attendance among people aged seven to eighteen years. In the case of Brazil, this means an emphasis in primary schooling.

Notwithstanding, there is a fair amount of evidence that the primary education conditionalities of Bolsa Família are both unnecessary and insufficient as a long-term strategy. They are unnecessary because primary school attendance in Brazil was already high at the time the program was scaled up. They are insufficient because primary school does not provide education enough to prevent poverty in the present and will hardly be capable of doing so in the future. Moreover, by targeting the poor only, the program misses a share of a special group under the risk of becoming poor due to low education, the children of the almost poor of today who may face downward social mobility.

At the time of implementation and expansion of TCCTs in Brazil, young children were already attending school, irrespectively of cash incentives. When Bolsa Família was scaled up, there was an increase in attendance of only two percentage points among the poor children and, of course, only part of this small improvement could be attributed to the program. Neither can it be argued that the TCCT contributed substantially to reduce the age-grade lag in primary schooling during the early 2000s, as most of this lag was reduced among children who started their school trajectories long before the program reached a large scale.

Indeed, a series of studies have shown the lack of relevant impact of Bolsa Família on education. Silveira-Neto (2010) uses Propensity Score Matching to estimate Bolsa Família's impact upon school attendance or enrollment and finds that the program increases attendance or enrollment by at most about two to three percentage points. Glewwe and Kassouf (2008) use the yearly School Census to estimate the effects of the program on enrollment, grade promotion, and drop-out. The limitation of their study is that the unit of analysis is the school and they know only if a school has at least one beneficiary enrolled, but not how many. Their results on grade promotion and drop-out, on the other hand, are so impressive as to be fishy: having a beneficiary in a school reduces drop-out probabilities by 31 percent and increases promotion probabilities by 53 percent. However, Camargo (2011) builds a variable that is the percentage of Bolsa Família students in each school. His results show no impacts at all upon attainment or achievement. Camargo has the better data and so we believe his results rather than Glewwe and Kassouf's.

Rogério Santarrosa (2011) and Heitor Pellegrina (2011) match individual standardized test data and Bolsa Família administrative records for the state of São Paulo. Both have access to socioeconomic and cognitive skill data, often for the same children, for 2007, 2008, and 2009. Using pooled regression analysis, fixed effects regression analysis, as well as various forms of matching and difference in difference estimators, Santarrosa concludes that the program has no significant effects upon cognitive skills. Using much of the same arsenal, Pellegrina concludes that it reduces dropout rates by something between two-tenths and one percentage point but also finds no impacts on learning. This should perhaps not be too much of a surprise because Bolsa Família requires the children to stay in school, but once there, they are taught by the same poorly paid and poorly motivated teachers, using the same limited pedagogical materials, and working in the same lousy installations in the same poor neighborhoods.

The existing evidence points to the fact that the primary education conditionality of Bolsa Família was innocuous because its objective had already been achieved by other policies. The Brazilian case only reflects a situation already found in other country studies: when school attendance is already high, the impact on education of demand-side incentives is weak.

Investments in primary education are positive but by themselves will not substantively reduce poverty in the country. Although primary education reduces the chances of being poor, this reduction is limited. The probability of being extremely poor for an adult with only primary education was 10 percent in 2009, which is less, but not much less, than the 17 percent observed for adults with less than primary. The reason for this is that the low wages of workers with primary education coexist with very high wages of people with secondary and tertiary education. Today, keeping the conditionalities as they are is hard to justify in terms of poverty eradication.

Nutrition and Health

One of the main reasons for fighting poverty is nutrition. Many poverty lines define poor as someone who does not have enough resources to buy food. Furthermore, Bolsa Família itself was assumed to be part of the Zero Hunger initiative implemented during the first term of Lula's government and later phased out. Lastly, there is ample evidence that poor nutritional conditions during childhood can hobble the productive capacity and well-being of individuals for the rest of their lives. However, the effectiveness of Bolsa Família, in this field, seems to be weak.

Segall-Corrêa et al. (2008) evaluate the program's impact upon perception of food insecurity, estimating a probit to show that every R$10 transferred by Bolsa Família reduces a perception index of food insecurity by approximately 8 percent. But these are subjective perceptions and not de facto measured child nutrition. Searching for more objective measures of impacts, Andrade, Chein, and Ribas (2007) compare anthropometric nutrition measures of children from six to sixty months in Bolsa Família families with those in comparable families with no benefit. The authors analyzed (1) height for age, (2) weight for height, (3) weight for age, and (4) body mass index for age. Separate analyses were undertaken in the northeast, north/center-west, and south/southeast regions of the country and for poor and very poor families separately. The results show no impacts at all. Children in families with a Bolsa Família benefit had the same nutritional profile as those with no benefit.

This is not a novel result. The nutrition effect for the Mexican program *Progresa* and its sucessor, *Oportunidades*, for example, is unclear. The nutritional component in Mexico was redesigned with nutritional supplements being distributed to the beneficiaries and, after that, positive impacts were observed on the height of children who were twelve to thirty-six months old (Behrman and Hoddinott 2005). However, there is no way of knowing whether this positive impact was due to the nutritional supplements given by the program or to the cash transfer itself.

A nutritional complementary program may be necessary for Bolsa Família. However, malnutrition in the conventional sense of the term—insufficiency of calories in the diet—is a residual problem in the country. Actually, Brazil is going through its nutritional transition and the excess, not the lack of calories in the diet of low-income families, is a growing problem. The impact of Bolsa Família in posttransition forms of malnutrition, such as vitamin deficiency and obesity, are yet to be evaluated.

Work

One of the criticisms repeated ad nauseam by political opponents of Bolsa Família is that it provides adverse labor market participation incentives to its beneficiaries. The criticism is that a means-tested transfer will create a labor disincentive, particularly in those extremely poor families that receive the fixed benefit, for which the only condition is being poor.

Microeconomic analysis of the labor participation effects of a transfer conditioned both upon school attendance and low income is not trivial. An unconditional transfer can increase or reduce labor supply because it involves only a wealth effect and does not change relative prices. When the benefit is means-tested, or conditioned to family income, relative prices change. For those labor participation decisions in which the program income eligibility line is not reached, the effect is the same as that of an unconditional transfer, but for labor decisions that go beyond the eligibility line, there is an unambiguous incentive towards reducing labor supply.

However—and this is important—the additional income of the cash transfers may have the same positive effects on self-employment as subsidized credit would have; and, as job search is a costly activity, the transfers may actually increase the chances of finding a job. This means even a means-tested transfer may increase employment. Finally, once child labor is considered, the inclusion of the education conditionality will lead to an unambiguous reduction of their labor supply, which may or may not affect the labor supply of the adult members but could conceivably increase it. In conclusion, economic theory does not shed much light on the existence of labor market disincentives, transforming it into a purely empirical issue.

There are at least seven good evaluations of the impact of Bolsa Família on labor supply. Given the importance of the issue, we will provide a brief overview of each. Ferro, Kassouf, and Levinson (2009) and Ferro and Nicollela (2007) apply probits and other regression methods to 2003 household survey data and find that child labor falls and adult labor rises. The effects, although significant, are small. Cardoso and Souza (2004) use the 2000 Census and Propensity Score Matching (PSM) to estimate the impact of those programs in existence in 2000 upon school attendance and child labor. Their results are a significant one percentage point reduction in female child labor and a one-half-point reduction in male child labor. Tavares (2010) also uses PSM to estimate the changes in the labor participation behavior of mothers and concludes that the benefit reduces the workweek from about 5 percent to 10 percent, which amounts to something between 0.8 and 1.7 hours. Statistically significant, but not exactly impressive.

Teixeira (2010) uses the intensity of treatment in a very creative paper. In other words she estimates if the size of the transfer has any impact upon the supply of hours in informal and formal activities. The effects are zero for both men and women in the formal labor market, but negative and significant for those in the informal sector. Despite their statistical significance the impacts are quite small. Foguel and Barros (2010) make a panel of municipalities covered by the Pnad Household Survey from 2001 to 2005. They find that a 10 percent increase in the proportion of beneficiaries in a municipality increases the female participation rate by a negligible 0.1 percent. For men, the effect is even smaller, only 0.05 percent. For hours worked the elasticity for women, in general, is about −0.01. Among men, there is no impact at all. Ribas and Soares (2011) use the same database, at a more disaggregated level (census tract), and with a slightly different methodology. They find that Bolsa Família increases informality and that rural women reduce by nine hours the time they dedicate to paid work. However, in the metropolitan areas their results point to a reduction in labor supply, not only labor market participation but also a reduction in labor force participation. This is slightly at odds with previous findings, although the effects are quite small and only borderline significant.

In conclusion, although the exact size and statistical significance of effects varies from study to study, they are always small. The possible exception is the number of hours worked by women with children. Given the importance of early childhood development, this is hardly a negative effect.

Fertility

Another specter from the dark that cash transfers to poor children always bring is that their parents will have more children in order to get more money. This fear led to the limits on the number of child benefits per family: originally three (now five) for children up to fifteen and another two for teenagers sixteen and seventeen. This can hardly be a regarded as a serious fear when Brazil's birthrate is below replacement level and we are looking at potentially serious demographic contraction effects in a few decades. A more justifiable ghost is that maybe young girls failing in school will be encouraged to get pregnant when they are still young, and this will severely handicap their education, frustrating the program's very objectives.

Rocha (2009) estimates Bolsa Família's impacts on fertility using a very clever approach in which he uses as an instrumental variable the program's three-child limit. He builds a treatment group composed of women with two children and a comparison group of those with three or more. Those with three or more have no monetary incentive from Bolsa Família to have more kids. The impact variable is whether the woman bore a child in the last twelve months. The results are that Bolsa Família has absolutely no effect whatsoever on fertility decisions.

In another study Alves and Cavenaghi (2011) use regression analysis of Bolsa Família beneficiaries in Recife and find no impact on fertility. Using more sophisticated propensity score-matching techniques and a national survey (the Pnad), Signorini and Queiroz (2011) find negative, but not statistically significant, effects of Bolsa Família on fertility. The authors argue that these effects are due to selection into the program that their model could not control, but a negative effect is quite plausible, because beneficiaries are required to go regularly to the doctor, where they will be exposed to family-planning methods.

CONCLUSIONS

Because of a series of local policies in parallel to the national ones, social protection is not completely homogeneous across the country. Yet, the national programs are the ones with larger scale and more impact on poverty and inequality. The term 'national', however, refers to the design, regulation, and funding of the programs. A good part of their execution and administration is decentralized and shared by different levels of government. Centralized funding ensures the programs will run irrespective of the fiscal capacity of lower levels of government; decentralized implementation, by its turn, requires administrative capacity at these lower levels.

The social protection system was developed with a strong corporatist bias and this has important implications for its distributive profile. After the end of the period of military rule in the 1980s, the prevalence of a social democratic tendency in Brazilian politics tried to reverse that bias. At some rate this play of forces created a polarized social protection system: contributory pensions are among the most concentrated incomes in the country and replicate preexisting inequalities, whereas social assistance and, to a minor extent, noncontributory and subsidized pensions are highly distributive. Because the weight of pensions is much higher than that of assistance transfers, the result is the state perpetuating inequalities via its social protection system.

Public pensions can be divided in two subsystems or regimes, RGPS for workers of the private sector and RPPS for the public sector. For analytical purposes, the pensions for workers of the private sector can be further subdivided into a group of noncontributory or highly subsidized benefits, all of which pay pensions at the floor equivalent to one legal minimum wage, and the more conventional contributory pensions. Social assistance at the national level consists of two programs: Bolsa Família, a conditional cash transfer targeted at the poor; and the unconditional Continuous Cash Transfer, BPC, targeted at low-income persons with disabilities or aged sixty-five or more years. Some subnational programs complement these two, but they have limited impact on inequality or poverty.

Around 28 percent of the Brazilian population lives in a household where at least one member is benefited by RGPS pensions. Less than 4 percent live in RPPS households, but the latter is so concentrated that its contribution to inequality is close to that of RGPS. The rules governing these subsystems are not the same, and the difference that affects inequality the most is the lack of a maximum value for the pensions of public workers. Only 1 percent of the population is benefited by pensions of higher values (above the RGPS limit), but about more than half of the contribution of RPPS to inequality is due to those pensions.

On the other hand, the imposition of a minimum benefit for noncontributory and subsidized pensions makes them egalitarian. In 2012 the majority of RGPS beneficiaries (65 percent) were receiving their pensions at the floor equivalent to the minimum wage. BPC assistance benefits, which have even stronger redistributive effects, not only use the same floor but are deliberately targeted at the poor using a fraction of the minimum wage as its eligibility line. Because the 1988 Constitution indexes the pension and BPC floors to the minimum wage, and because there is a law determining that every year this wage must be adjusted for inflation and receive a real increase proportional to the growth of GDP, the role of the minimum wage in the Brazilian social protection is and will be crucial for the next decade or so.

As BPC, Bolsa Família is very well targeted. But differently from the former, its transfers are not bounded by the minimum wage floor. In fact, Bolsa Família transfers much less money per family than the nonconditional protection programs. Because it was scaled up with excellent targeting, it had an important role in the fall of inequality observed in Brazil during the 2000s and, of course, in alleviating poverty, but it still is insufficient for eradicating poverty. Given its excellent coverage of the poor, rises on the values transferred would increase substantially its effects on poverty, still at a reduced fiscal cost.

Bolsa Família is very important as an antipoverty policy. Besides the impacts the transfers currently have, the program has set the institutional infrastructure needed for an eventual expansion and created the political space for more distributive social protection in the country. All conditionalities of the program, however, seem to be ineffective. A series of evaluations have shown the lack of relevant impacts of Bolsa Família on education, nutrition, and health. Neither does it seem to have impacts that could be considered negative such as reductions in labor market participation or increases in fertility.

REFERENCES

Alves, José Eustáquio, and Cavenaghi, Suzana. 2011. "O Programa Bolsa Família e políticas públicas: saúde reprodutiva e pobreza na cidade do Recife." Paper presented at the IX Encontro Nacional da Associação Brasileira de Estudos Regionais e Urbanos.

Andrade, Mônica V., Chein, Flávia and Ribas, Rafael P. 2007. *Políticas de Transferência de Renda e Condição Nutricional de Crianças: Uma Avaliação do Bolsa Família.* CEDEPLAR Discussion text no. 312. Belo Horizonte: Cedeplar.

Behrman, Jere R. and Hoddinott, John. 2005. "Programme Evaluation with Unobserved Heterogeneity and Selective Implementation: The Mexican PROGRESA Impact on Child Nutrition" in *Oxford Bulletin of Economics and Statistics.* Volume 67, Issue 4, pages 547–569, August 2005.

Camargo, P. C. 2011. *Uma analise do efeito do Programa Bolsa Familia sobre o desempenho medio das escolas brasileiras.* Dissertation (Master's in Economics)—Programa de Pos-Graduacao em Economia, Universidade de Sao Paulo, Ribeirao Preto, 2011. 58p.

Cardoso, Eliana.; Souza, André. P. 2004. *The impact of cash transfers on child labor and school attendance in Brazil.* Nashville: Department of Economics, Vanderbilt University. (Working Paper 04-W07). 51 p.

Ferro, Andrea R., Kassouf, Ana Lúcia, and Levison, Deborah. 2009. "The Impact of Conditional Cash Transfer Programs on Household Work Decisions in Brazil." In *Anais do XXXVII Encontro Nacional de Economia* (Proceedings of the 37th Brazilian Economics Meeting) Salvador. Anpec. Available at http://econpapers.repec.org/paper/anpen2009/208.htm

Ferro, Andrea R., and Nicollela, Alexandre C. 2007. "The Impact of Conditional Cash Transfers on Household Work Decisions in Brazil." Paper presented at the IZA/World Bank Conference on Employment and Development.

Foguel, Miguel N., and Barros, Ricardo P.. 2010. "The Effects of Conditional Cash Transfer Programmes on Adult Labour Supply: An Empirical Analysis Using a Time-Series-Cross-Section Sample of Brazilian Municipalities" *Estudos Economicos* 40 (2): 259–93.

Glewwe, Paul, and Kassouf, Ana Lúcia. 2008. "The Impact of the Bolsa Escola/Familia Conditional Cash Transfer Program on Enrollment, Grade Promotion and Drop out Rates in Brazil." In *Anais do XXXVIII Encontro Nacional de Economia,* Salvador: Anpec. Available at http://www.anpec.org.br/encontro2008/artigos/200807211140170-.pdf

Hoffmann, Rodolfo. 2003. "Aposentadorias e pensões e a desigualdade da distribuição da renda no Brasil." *Econômica* 5 (1): 135–44.

———. 2007. "Transferências de Renda e Redução da Desigualde no Brasil e em Cinco Regiões, entre 1997 e 2005." In *Desigualdade de renda no Brasil: uma análise da queda recente,* Vol. 2, edited by Barros, Ricardo P., Foguel, Miguel N., and Ulyssea, Gabriel, 17—40. Brasília: IPEA.

———. 2009. "Desigualdade da distribuição da renda no Brasil: a contribuição de aposentadorias e pensões e de outras parcelas do rendimento domiciliar per capita." *Economia e Sociedade* 18 (1): 213–31.

Marques, Rosa Maria, and Euzéby, Alain. 2005. "Um regime único de aposenta-doria no Brasil: pontos para reflexão." *Nova Economia* 15 (3): 11–29.

Medeiros, Marcelo; Britto, Tatiana, and Soares, Fabio V. 2007. "Transferência de renda no Brasil." *Novos Estudos—CEBRAP 5–21*.

Melo, C. R, and F. Anastasia. 2005. "A reforma da previdência em dois tempos." *Dados* 48 (2): 301–32.

MPS. 2004. "Boletim Estatístico da Previdência Social," vol. 9, no. 03.

———. 2012. "Boletim Estatístico da Previdência Social, vol. 17, no. 3.

Osorio, Rafael G., Soares, Sergei D., Souza, Pedro H. G. F. 2011. *Erradicar a Pobreza Extrema: Um Objetivo ao Alcance do Brasil.* Working paper Ipea n. 1619. Brasília: 58 p.

Osorio, Rafael Guerreiro, Pedro Herculano Guimarães Ferreira Souza, Sergei Soares, and Luis Felipe Batista Oliveira. 2011. *Perfil da pobreza no Brasil e sua evolução no período 2004–2009.* Working paper Ipea n. 1647. Brasília: 50 p.

Pellegrina, Heitor. 2011. "Impactos de Curto Prazo do Programa Bolsa Família sobre o Abandono e o Desempenho Escolar do Alunado Paulista." Master's diss., University of São Paulo. http://www.teses.usp.br/teses/disponiveis/12/12138/tde-26092011-165149/pt-br.php (accessed 12–04–2013).

Ribas, Rafael Perez, and Soares, Fabio V. 2011. *Is the Effect of Conditional Trans-fers on Labor Supply Negligible Everywhere?* Mimeo. Electronic copy available at: http://ssrn.com/abstract=1728287 . Brasília: 45 p

Soares, Fabio V.Rocha, Romero. 2009. "Programas Condicionais de Transferên-cia de Renda e Fecundidade: Evidências do Bolsa Família." Paper presented at Sociedade Brasileira de Econometria.

Santarrosa, Rogério. 2011. "Impacto das Transferências Condicionadas de Renda sobre a Proficiência dos alunos do Ensino Fundamental no Brasil." Master's dis-sertation., Fundação Getúlio Vargas— Escola de Economia de São Paulo.

Segall-Corrêa, Ana Maria; Marin-Leon, Leticia, Helito, Hugo, Ribas, Rafael P., Santos, Leonor Maria P., and Paes-Souza, Rômulo. 2008. "Transferência de Renda e Segurança Alimentar no Brasil: Análise dos Dados Nacionais." *Revista de Nutrição* 21 (supplement): 39–51.

Signorini, Bruna, and Queiroz, Bernardo. 2011. *The Impact of Bolsa Família Pro-gram in the Beneficiary Fertility.* Cedeplar Texto para Discussão no. 439. http://cedeplar.ufmg.br/pesquisas/td/TD%20439.pdf (accessed 12–04–2013). Belo Horizonte: Cedeplar .

Silveira-Neto, Raul M.. 2010. "Impacto do Programa Bolsa Família Sobre a Freqüência à Escola: Estimativas a Partir de Informações da Pesquisa Nacional por Amostra de Domicílios (PNAD)." In *Bolsa Família 2003–2010: Avanços e Desafios*, vol. 2, edited by Castro, Jorge A.and Modesto, Lúcia. 53–71. Brasília: IPEA.

Soares, Fabio V. 2011. "Bolsa Família: A Review." *Economic & Political Weekly* 46 (21): 55–60.

Soares, Fabio V., Ribas, Rafael P., andOsorio, Rafael G. 2007. *Evaluating the Impact of Brazil's Bolsa Familia: Cash Transfer Programmes in Comparative Perspective.* International Policy Centre for Inclusive Growth. http://ideas.repec. org/p/ipc/pubipc/296686.html (accessed October 2, 2009).

Soares, Fabio V., Ribas, Rafael P.and Osorio, Rafael G. 2010. "Evaluating the Impact of Brazil's Bolsa Família: Cash Transfer Programs in Comparative Per-spective." *Latin American Research Review* 45 (2): 173–90.

Soares, S. 2006. "Análise de bem-estar e decomposição por fatores da queda na desigualdade entre 1995 e 2004." *Econômica* 8 (1): 83–115.

Soares, Sergei. 2009. "Volatilidade de renda e a cobertura do Programa Bolsa Família." *IPEA, Texto para Discussão* (1459).

Soares, Sergei, Osorio, Rafael G., Soares, Fabio V., Medeiros, Marcelo, and Zepeda, Eduardo. 2009. "Conditional Cash Transfers in Brazil, Chile and Mexico: Impacts upon Inequality." Special issue. *Estudios Económicos* 207–24.

Soares, Sergei; Souza, Pedro H. G. F.; Osorio, Rafael G. and Silveira, Fernando G. 2010. "Os impactos do benefício do Programa Bolsa Família sobre a desigualdade e a pobreza." In *Bolsa Família 2003–2010: avanços e desafios*, vol. 2, edited by Jorge Abrahão Castro and Lúcia Modesto, 366. Brasília: IPEA.

Tavares, Priscilla A.. 2010. "Efeito do Programa Bolsa Família sobre a oferta de trabalho das mães." *Economia e Sociedade* 19 (3): 613–35.

Teixeira, Clarissa G.. 2010. "A Heterogeneity Analysis of the Bolsa Família Programme Effect on Men and Women's Work Supply." IPC Working Paper no. 61. Brasília: IPC.

3 Growth and Social Policies
Towards Inclusive Development

Jan Nederveen Pieterse[1]

INTRODUCTION

Well into the twenty-first century, many developing countries are more prosperous than in past decades and engage in large-scale social policies. Welfare policies that were absent or thin in the past are taking on a significant scale. However, they are often disconnected from economic policies. The ministries of economics and of social affairs don't speak to each other, or when they do they don't speak the same language. Economists have mostly been trained in neoclassical economics, and in some cases in Chicago school supply-side economics, whereas social affairs ministries speak the language of social cohesion and political stability. Thus, the logics of accumulation and welfare, of growth and social policy don't connect. This policy schizophrenia is not occasional. It reflects long-standing divides between economic and social spheres. It mirrors the long-standing disconnect on a world scale between the international financial institutions, based in Washington, and the UN institutions.

Social policies played a central role as part of the Keynesian consensus of the postwar era; they were marginalized during the period of neoliberal globalization; and they resurfaced to play a new role in the wake of the 2008 crisis. Social policies are further refracted by differences between developing and developed economies, and between social market and neoliberal economies.

This discussion first takes up the question of the relationship between social policies and economic growth: are social policies a benevolent afterthought of growth, or are they part of growth strategies? In addition, is redistribution a viable policy framework? The chapter then turns to the impact of globalization, information technology, financialization, and export-led growth. Patterns have been changing again in the emerging multipolar world and in the wake of the 2008 crisis. Given slowing world trade and high dependence on exports in many developing countries, a social turn in growth strategies can serve as a key component in moving away from export-led growth toward domestic demand-led growth. The scope of the discussion is global, with an emphasis on developing countries and emerging societies.

GROWTH AND REDISTRIBUTION

Growth and social policies are often viewed either as unfolding on separate tracks or in a sequence of growth first, redistribution after. The idea is that sharing without growth would only produce shared poverty. Whereas this may generally make sense, it is also too general to hold much water—as if any growth is welcome and the quality of growth doesn't matter. In effect this recycles the idea of 'growth above all' and doesn't count social and environmental cost. Besides, this sequence in effect means trickle down, and it entrenches interest groups and policies that create their own path dependence. Most important, it glosses over the importance of human capital as a key factor in productivity and growth.

'Redistribution and growth' continues to be argued as a poverty reduction policy that is more effective than growth alone (Dagdeviren, Van der Hoeven, and Weeks 2002), which is true but is also an easy argument. Is growth and redistribution a formula we can go back to, or does it belong to a bygone era? The 1970s growth and redistribution literature (Chenery et al. 1974) came during the waning years of the Keynesian consensus. As a concept and policy framework, redistribution carries several limitations. Redistribution assumes a stable political center and effective fiscal and revenue raising policies. Let's note that in the United States none of these conditions apply: the political center isn't stable because of elections every two and four years; fiscal policies aren't effective because corporate lobbyists and lawyers create or find tax loopholes; and raising revenue is difficult with major political forces opposing tax increases.

Redistribution further assumes a stable social contract, a social consensus, which isn't available in societies that are deeply divided—in societies with structural conditions of radical inequality such as India, Pakistan, and much of Southeast Asia; and in heterogeneous immigrant societies such as the US. It also poses a problem in societies where immigration has increased amid economic constraints, as in much of Europe. Thus, precisely where redistribution is most needed, the social basis and political coalition to achieve it is least likely to materialize. In the US 'redistribution' is a nonstarter and an ugly word that smacks of 'big government' and socialism. What has taken place in the US since the 1970s is redistribution-in-reverse, a vast reconcentration of wealth and power, undoing the reforms of the New Deal and returning the country to the wealth disparities of the 1920s. In continental Europe, redistribution is ordinary, but immigration exposes the limits of the social contract. In Mediterranean Europe welfare has taken on forms of lax state patronage, and the current crisis in the eurozone exposes the design problems of the European Union. In developing countries, large-scale welfare policies have been a new trend, and how they relate to growth is contentious, which is the focus of this discussion.

Redistribution assumes effective distribution policies and capable local government, which is problematic in many developing countries, China and

India included. Social transfers may be subject to elite capture at the local level. As a discourse, redistribution is elastic, is subject to narrow or wide interpretations, holds different meanings for different policy makers, and is therefore unstable as a policy framework. Furthermore, redistribution per se ignores or holds constant the overall growth paradigm. Growth and redistribution are viewed as separate compartments, or alternatively, the implicit assumption is a Keynesian policy framework. In addition, redistribution ignores macroeconomic dynamics. It is subject to imponderable economic fluctuations and implies a Keynesian 'national economy' bias, i.e., a lesser degree of globalization than exists under conditions of complex interdependence.

The slumps of the 1970s and 1980s turned the tide, and at the World Bank the Washington Consensus outflanked the redistribution and growth approach. In development studies, the human development approach took over the legacy of growth and redistribution (Murphy 2006). In the 1990s debates followed on trade-offs between growth and equity (Nederveen Pieterse 2010a). The human development approach argues not for growth but for quality growth, and not for redistribution but for capacitation.

SOCIAL POLICIES

Thomas Pogge (1998) refers to conditions of 'radical inequality': those at the bottom are very badly off in absolute terms; they are also very badly off in relative terms; and the inequality is persistent, pervasive, and avoidable. In several countries some of the conditions of radical inequality have begun to unravel. With social forces and newly empowered strata emerging in emerging markets, they have become emerging societies. Political demands are stronger, government coffers fuller, and social policies more ambitious—but what is the political economy of the ongoing transformations, and how do they fit into the wider economic and political equations?

Social reforms such as cash transfers (as in Mexico's Oportunidades and Brazil's Bolsa Família), work programs (such as India's NREGA), microcredit (as in Grameen), and social provisions (such as health care and pension schemes in Thailand and child care support in Nepal) share the limitations of redistribution policies. As long as they are conceived as set *apart* from the logics of growth itself, they are vulnerable to political vicissitudes and economic fluctuations.

The July 2011 elections in Thailand illustrate the dilemma. The Pheu Thai party led by Yingluck Shinawatra (the sister of the deposed and exiled former prime minister Thaksin Shinawatra) won a landslide victory on a program of major social benefits,[2] which is as populist as Thaksin's policies had been. Economists and business leaders cautioned that these policies would lead to inflation, higher interest rates, and higher costs for private companies (*Bangkok Post*, "Election 2011," July 5 2011, p. 6). The election victory represented a renegotiation of Thailand's social contract against the backdrop of political crisis with "warring political factions, five years of street protests and violent military crackdowns." This includes the long-neglected countryside, especially in the northeast. A report notes: "Once passive and fatalistic, villagers are now better educated, more mobile, less deferential and ultimately more politically demanding. . . . The old social contract, whereby power flowed from Bangkok and the political establishment could count on quiet acquiescence in the Thai countryside, has broken down. . . . Villagers describe a sort of democratic awakening in recent years and say they are no longer willing to accept a Bangkok-knows-best patriarchal system" (Fuller 2011). Thus, structural changes in the countryside and the "transformation from 'peasants to cosmopolitan villagers'" (quoted in Fuller) and mounting social pressure are fundamental to the ongoing political changes. According to the Pheu Thai party, the social policies will stimulate domestic consumption and help lead the country away from depending on exports. Its Vision 2020 refers to the ASEAN Vision 2020. Missing in the mix, so far, are land reform and broadening access to education.

The general quandary is whether social policies are redistributive trickle down or whether they are conceived as *part* of the growth model. Are social policies a bonus to poorer strata for overall growth, a low-cost way of buying social peace and cutting crime (as in Brazil since Presidents F. H. Cardoso and Lula)? Are social policies market-friendly (such as credit schemes that enable the poor to buy on credit), or are they embedded in expanded worker rights such as collective bargaining (as in Germany's coordinated social market economy; Vaut 2012)? Are they occasional redistribution or do they reflect a social perspective on growth and a different growth path? Obviously these are as much political questions as economic questions.

Elite patronage and charity are social policies that are disarticulated from growth policies. The emblematic case is the nineteenth-century soup kitchen. They are typically short term, depend on market and political fluctuations, and have a demobilizing effect. Some social policies may serve as repair of damage done, as in Spain, Portugal, and Greece after the fall of dictatorships, or in the 'new South Africa' after 1994. As reparation policies they fall short of a new social contract. Some forms of charity stem from entitlements to food staples and food security that may go back to feudal lordly duties, and over time they may transform into rights. If social policies are entitlements and based on rights, as in India's right-to-work program, they go beyond trickle down. India's NREGA program changes the village power structure and the sway of caste rule in the panchayats because it provides a revenue source outside the village. In sum, to be genuinely effective and sustainable social reforms should be part of the overall growth model and engage macroeconomic imbalances.

The main ways of integrating growth and social policies are demand-side and supply-side approaches. On the demand side, production without consumption and steep social inequality and wealth concentration is not sustainable. Growing inequality, as in Marx's pauperization thesis, under-cuts mass demand, the classic quandary emphasized by Keynes, Galbraith, and recently by Robert Reich and Paul Krugman.

On the supply side, the key variable is capacitation. Social policies enhance broad social participation in growth; growing human capabili-ties—sustained by education, health, and housing policies—boost produc-tivity and employability, widen the tax base, and thus establish a virtuous circle. In the Nordic countries this is the productivist approach to wel-fare, or social productivism, as in Gunnar Myrdal's classic work (1944). In Germany it is the combination of welfare policies, apprenticeship, codeter-mination in shop floor affairs and company boardrooms, and the partner-ship of government, employers, and trade unions. In Rhineland capitalism, capacitation and social inclusion likewise go together (Albert 1993). In East Asia similar policy combinations have inspired the human develop-ment approach. This goes much further than redistribution; it involves restructuring growth models and changing the political equation. It also goes further than human development. It isn't just a matter of individual attributes as in the Human Development Index but concerns building social institutions, so it is social development. As Wilkinson and Pickett (2009) show in their impressive study *Why More Equal Societies Almost Always Do Better*, all of society benefits from equality, also the rich, with less crime, less disease, greater security, and social stability. The recent Spence Commission's case for shared growth and inclusive development takes this to a further level (Commission on Growth 2010). Table 3.1 sums up ways of articulating social policies and growth.

Some of these relations have broken down in the setting of contemporary accelerated globalization and IT growth. In developed countries they have been short-circuited by post-Fordism, offshoring and outsourcing, and the

Table 3.1   Articulating Social Policies and Growth

Demand Side	Develop social demand and sustain domestic market
Supply Side	Capacitation: enhance broad social par-ticipation in growth; education, health, and housing policies sustain growing capabilities, boost employability, and widen the tax base
	Social investment: in productivity and social cohesion; education and empow-erment of women, minorities
	Social development: capacity and institu-tion building

creation of low-paying jobs in the service sector. In the wake of the 2008 crisis, the issue is how to rework and reconceptualize social policies.

Another side of the story is the privatization of social security. Chile's private pension reform in 1981 led the way, followed by World Bank reports recommending the privatization of pensions and health care. A recent report observes:

> Far from increasing efficiency, the reforms have proved costly and have drained public resources through lavish tax incentives and significant administrative and regulatory expenses. In Chile, the private pensions system absorbs around a third of the overall government budget and 42 per cent of public social expenditure. The administrative costs associated with private health insurance have been estimated to be up to ten times higher than the administration costs of social insurance. There has been a failure to increase coverage, as only those who can afford to pay premiums can benefit from private schemes and high risk individuals are excluded. Women, who make up a large proportion of informal workers and the poor, often receive significantly lower benefits, and are doubly hit in the face of declining public expenditure on social security. . . . many private pension fund management companies are in the hands of foreign financial conglomerates. Chile's largest private pension manager, Provida, with $36.1 billion under management, is owned by Spain's largest financial institution BBVA. Between 1981 and 2006, Chilean workers contributed approximately $50 billion from their salaries towards the private pension schemes, of which private pension managers and related insurance companies kept one third as commissions and profit. (Sumaria 2010: 2, 5)

The privatization of social security, abetted by the World Bank and IMF, has been one of the avenues of financialization in developing countries, many of which have implemented more drastic privatization reforms than developed countries. As Sheena Sumaria notes, "Social security privatization is not necessarily about the retreat of the state from social protection, but about the state's transformation" (2010, 5).

GLOBALIZATION, IT, FINANCIALIZATION, AND EXPORT-LED GROWTH

The Keynesian consensus broke down at the intersection of the slumps of the 1970s and 1980s, accelerated globalization, technological change, and financialization. The Fordist approach matched productivity growth with wage increases plus inflation. In the 1980s technological changes enabled post-Fordism, flexible production, automation, containerization of ocean transport, and 24/7 (24 hours 7 days a week) global finance, and hence the further interweaving of firms and economies across national boundaries. Global competition, the need to invest in technology and marketing, and gaining global market share reinforced the trend towards offshoring to zones with lower labor cost. About the same time, Eastern Europe and China opening up to international markets added a vast pool of labor to the global labor market (Prestowitz 2005).

Together these changes enabled a major shift in bargaining power from labor to capital along with a different understanding of growth, led by capital, hence the rise of shareholder capitalism. An additional factor is that American society is steeped in a culture in which business occupies a larger place than in past hegemons and in other societies. It includes within it a vast zone, the American South, where tax, labor, and regulation standards have lagged far behind the rest of the country. Dixie capitalism enabled the neoliberal turn in the US (Nederveen Pieterse 2004). The international financial institutions based in Washington, the IMF and World Bank, became instruments of this outlook, the Washington Consensus.

In the 1980s the Washington institutions promoted export-led growth as the leading development model, combined with trade liberalization, deregulation, and privatization, even though the success stories of this model, the Asian Tiger economies, all involved active government intervention. Export-led growth along with investment from American companies enabled the rapid industrialization of Asian economies. The Cold War and American wars in the region (Korean War, Vietnam War) also provided stimuli. For Korea and Taiwan the proximity of Japan mattered as well. Thus, export-led growth and industrialization in developing countries has been a mirror image of deindustrialization in OECD countries.

What ensued is a complex interdependence of Pacific economies. In brief, manufacturing and service jobs lost in the US led to rising wages in Asia. In the US productivity has been rising steeply and corporate profits rose, but wages remained stagnant broadly since the 1970s. Profits, the Dow Jones, and CEO remuneration are up because American corporations reap high yields from rising productivity and from offshoring. Cheap Asian imports compensate for stagnant American wages, but over time rising wages and the skills squeeze in emerging economies will raise the cost of imports and will make offshoring to such areas marginally less attractive.

Since the 1980s this growth model was sustained by the US as driver of the world economy with private consumption as the main engine (rising from 64 percent in 1980 to 72 percent of US real GDP in 2007). With wages remaining stagnant, consumption was sustained by longer working hours, double-earner households, and credit expansion (deferred payments, credit cards, home equity financing), made possible by low interest rates and external borrowing. Credit expansion fueled the financialization of the economy; financial services became the largest sector of the US economy, employing 20 percent of the workforce and generating 40 percent of corporate profits. This further deepened inequality, with pay rates in finance much higher than in other sectors (Schmitt 2012). Deregulation combined with new financial instruments—some arcane (credit default swaps, derivatives, securitization) and some fraudulent (subprime mortgages)—and growing white-collar crime contributed to financial instability and crisis, peaking with the fall of Lehman Brothers in 2008 and ongoing.

'Globalization', then, is shorthand for a package deal of concurrent changes, a vortex of interacting forces. In the 1990s 'globalization' became a buzzword to cut government intervention on the grounds of competition and capital flight to low-cost and low-regulation zones. Globaloney globalization (or globalization of the managerial variety) intertwined with turbo capitalism helped transform social market and stakeholder capitalism into 'no-nonsense capitalism' (a term used in the Netherlands at the time to refer to lean capitalism, stripped of generous social benefits). However, the *form* of globalization during a particular phase is not the same as the *trend* of globalization. Second, the momentum of globalization is more complex and points in more directions than just the course orchestrated by the hegemon. Third, hegemons don't last. Fourth, in the wake of the 2008 crisis and the global imbalances it reveals, the neoliberal turn, although it is not gone, is over its peak as ideology and difficult to sustain institutionally.

TIPPING POINTS

These developments are now at a crossroads because of economic decline in the OECD, and besides, because it produces social forces, as in Thailand, which cannot be contained within this social constellation. These relations are unstable for American trade, and current account deficits cannot rise indefinitely. Tipping points include the limits to American purchasing power (in view of stagnant wages, steep inequality, and crisis), debt, and the unstable dollar.

At this stage the American model of import and borrow and the Chinese model of export and lend are both unsustainable. "If the import-and-consume business model is dead, so too is export-and-save" (*Financial Times* editorial, April 16, 2009, p. 8). According to Thomas Palley, "the possibility of global development via export-led growth is now exhausted." Key problems he notes are waning consumer markets in developed economies; emerging markets' exports hinder the recovery of industrialized economies; emerging markets' exports crowd out the exports of other emerging markets; increasing South-South competition; declining prices of manufactured goods; and the ability of multinational corporations to shift production to lower cost countries (Palley 2011: 4–5).

In the US a key problem is private underinvestment. The rise of outward investment in the 1980s coincided with a peak period of American hegemony. American companies' offshoring and investing outward was facilitated by the Department of Commerce and the Export-Import Bank as part of the projection of American hegemony. Whereas in Germany, the European Union, and Japan offshoring was generally *balanced* by *inward* investment and innovation, in the US deindustrialization and financialization went much further (also in the UK because of reliance on the London City as financial center). Hence in the course of two decades US manufacturing exports became imports, at the expense of American jobs and growing trade and external deficits. When at this stage the US attempts economic recovery, lagging inward investment and innovation undermine the global competitiveness of American products, except in a few sectors (such as military industries, software, and pharmaceuticals); a case in point is green technologies in which American companies lag behind in every sphere—wind turbines, solar panels, biofuels, and energy efficiency (Nederveen Pieterse 2010b).

Exports substituted for domestic demand also in Europe: "The solution to the problem of effective demand is seen as lying above all in a positive trade balance. . . . This outlook on the part of capitalist institutions and firms . . . relegates the domestic level of employment and of wages to a subordinate role compared with external expansion. Profits accruing from net exports reduce firms' dependence on a relatively small or slow-growing domestic market, and Europe's surplus countries are well aware that were it not for their export strategy domestic investment, profits and employment would be lower" (Bellefiore, Garibaldo, and Halevi 2011: 120).

Decoupling (emerging markets becoming independent from Western demand) has not materialized; weak growth in all the leading economies together poses a problem. Alternative markets for Asian and emerging markets' products—in Asia (ASEAN+6), East-South trade, and domestic markets—are taking shape, but at a slower pace than OECD demand is falling. Thus, at this stage mercantilism poses a fundamental problem in the world economy. If all countries rely on export-led growth—who imports? Export-led growth together with steep inequality feeds the overall dynamics of overproduction-underconsumption.

Part of the quandary is the dynamics of financialization unfolding on a global scale. Following the Asian crisis (1997–98), developing countries accumulated financial reserves as a buffer against turbulence. Several flows fuel financialization on a global scale: credit expansion in import-dependent countries (especially the US); savings, trade, and financial surplus in exporting countries; developing countries' buffers to ward off financial turbulence; developed economies' crisis management with bailouts, stimulus, and 'quantitative easing'(QE). QE also works as a policy of exporting inflation that triggers higher commodity and food prices and increasing inflation risks in emerging markets. In Brazil the policies of the US Federal Reserve have been dubbed a 'currency war', causing an inflow of foreign capital seeking returns, appreciation of the Real, and a high interest rate.

Financial investments in emerging markets' industries enable Western pension funds and institutional investors to secure financial returns that sustain the income of pensioners. Conversely, exporting economies' surplus and savings lent to the US in Treasury bond purchases contribute to the financialization of the world economy. The Chinese lending their hard-earned dollars to the US contribute to ballooning deficits in US dollars that are worth less. Thus, Chinese labor subsidizes the American economy in several ways: by providing low-wage labor, cheap products, Treasury purchases that help keep US interest rates low, and returns on investments that keep American pensioners going. Capital controls, safeguards against inflation and property bubbles, and restrictions on international finance are high on the agenda of emerging societies. At the macroeconomic level, Dilma Rousseff, president of Brazil, advocates "further regulation of the financial system, to minimise the possibility of new crises; reduced levels of leverage. We must proceed with the reform of multilateral financial institutions, increasing participation of emerging countries that now bear primary responsibility for global economic growth" (2011).

WELCOME TO THE MULTIPOLAR WORLD: SOCIAL POLICIES REVISITED

The 2008 crisis ushers in a new phase in the interaction of capitalisms. The Washington Consensus survived the crises of the 1990s in tatters. In the wake of the 2008 crisis, we leave what remains of it behind and the question is, for what—for the Rhineland model, the German model, the East Asian model, or a 'Southern consensus' (combining the Beijing, Delhi, and Brasília consensus)? In the wake of the 2008 crisis, the question of social policies and growth takes on different equations in each of the major zones of capitalism.

Emerging markets now drive the world economy, and East Asia is widely regarded as the main 'winner' in contemporary globalization. During the period 2000–2007, which has seen the fastest growth of world trade in history, Asian developing countries' ratio of exports to GDP rose from 36 percent to 47 percent in 2007. Thus, East Asia is in the lead—but tethered to a postbubble world economy.

East Asian developing countries' scripts include engaging in global competition and emphasizing science, innovation, and design; and shifting gear from price competition to quality and brand competition, and from industry to services and from tradable to nontradable goods. Going green and commodity-lite is another challenge. Replacing exports to the US and EU with regional and Global South demand is a long haul. At this juncture, the risks for emerging markets and developing Asia are betting on export-led growth when global trade slows, entering global finance when Anglo-American megabanks rule, and pursuing peaks while neglecting the social base. For emerging societies, then, the trade-off is between global competitiveness and building the domestic market; too far a tilt in either direction jeopardizes their balance. It is a matter of balancing peaks and valleys. Investing in peaks—science, technology, design, finance—is necessary to sustain global competition; investing in valleys—in social security, broadening access to education, reducing the need for savings, investing in agriculture, and pro-consumption policies—is necessary to build domestic demand, reduce reliance on exports, and sustain and deepen democracy. Cutting dependence on exports and shifting gear from supply-driven growth to demand-led growth, driven by domestic consumption, is essential.

This is where articulating growth models and social policies comes in (an overview of social policies in developing countries is Burchardt, Tittor, and Weinmann 2012). In emerging societies social policies sustain broad effective demand and sync with abandoning export-led growth. A social protection floor (as advocated by the ILO; Wodsak 2012) also tempers the effects of economic downturns. However, the continuing emphasis on global competition is an incentive to keep wages and prices low. Thus, in South Korea the central social issue is the growing dualization between regular and irregular workers, with the latter receiving much lower pay and lesser labor conditions (Lee and Jeong 2011).

Among the BRICs and emerging societies, China is most advanced in merging social policies with reorienting its growth path. Hu Jintao's 'scientific outlook on development' and Wen Jiabao's 'five imbalances' of the Chinese economy set the stage for the twelfth five-year plan, 2011–15. This aims at building broad social safety, reducing the need for savings, and thus boosting domestic consumption, reorienting the economy away from export dependence (Li 2012; Roach 2009). The aim is to eventually balance China's external accounts so China would import as much as it exports. In Wen Jiabao's (2011) words, "We have made breakthroughs in building a social security system covering urban and rural areas. We have introduced a rural old-age insurance scheme which will cover 60 percent of counties in China this year. The basic urban medical insurance scheme and rural cooperative medical care scheme now cover more than 90 percent of the population." At issue is 'shifting the development model'. For the world's second-largest economy, much is at stake. Implementing this may slow growth rates and runs counter to powerful export interests, so it is a long process. The 18th Party Congress of November 2012 reaffirms the commitment to dealing with China's economic imbalances (Hille 2012).

Turning to developed countries, they have been on a technological plateau for some time, as Cowen (2011) argues. The 1990s were a time of economic stagnation that was papered over by financial expansion, a period marked by overleveraging and the steep growth of inequality. The dikes broke in the 2008 crisis. The bailouts socialized bank debt, ushering in phase two of the crisis, the sovereign debt crisis. Instead of regulation there has been consolidation of the financial sector, leaving six megabanks standing in the US.

In an age of deleveraging, when economies contract politics stumbles. In the US this means political gridlock and a split Congress, as in the debt ceiling and budget controversies. It may take several electoral cycles for

Americans to make up their mind where they want to go, with laissez-faire or with government coordination, with the Austrians or the Keynesians.

In the eurozone deleveraging exposes the architectural design problems of the EU (monetary union but no alignment of fiscal policies; a cumbersome voting system and democratic deficits). It exposes the tensions between Europe's disparate economies, with southwest Europe investing in industry, technology, and infrastructure, Mediterranean Europe investing in real estate and speculative property (Spain) or extending state patronage (Greece), whereas Wild West frontiers of finance emerged in Iceland and Ireland.

Thomas Palley proposes a domestic-demand led strategy—along with social safety nets, raising wages, improved labor protection, and collective bargaining by unions; public infrastructure investment; investing in health care and education; and reforms to make taxes more progressive (2011: 5–6). But this is hardly feasible in economies in which neoliberalism is institutionally entrenched. Etzioni (2011) observes that moral capital and political capital are in limited supply, more limited than many imagine, so he argues for 'policy minimalism' as a virtue. In some countries it may rather be a necessity. The trilemmas of balancing employment, fiscal soundness, and equality (Iverson and Wren 1998) work out differently in different capitalisms. Neoliberal, social democratic, and Christian democratic societies face these trilemmas in different ways (Im 2007).

Inequality has been growing worldwide and particularly steeply in the US and UK. In the US and UK, growth led by the financial sector prompts luxury consumption whereas the Main Street economy is slowly crumbling, producing an hour-glass society. Worldwide some 500 billionaires own as much as half the world population. As extreme capitalism produces radical inequality, it gradually undermines its own sustainability. In the US giant corporations continue offshoring and outsourcing; financialization continues and megabanks are the latest phase of American hegemony. The formula of billions for banks and austerity for people has reached a breaking point; witness the Occupy Wall Street (OWS) movement, riots in England and Greece, protests in Spain and Italy in response to draconian austerity, and the tent movement in Israel. The OWS movement is part of the globalization of anger; impunity for white-collar crime and financial corruption has reached a tipping point. The bottom line is that countries that don't invest in the future decline.

Social market economies that continue to invest in manufacturing, technology, infrastructure, and education may plow through the crisis, which may apply to Germany and Nordic Europe (assuming the problems of the eurozone can be managed) and Japan. They may experience lower rates of growth, but with graying populations growth is less important.

In conclusion, the Commission on Growth and Development's case for shared growth as the norm for sustainable growth and development is valid across the spectrum from developing to developed societies, and to establish shared growth, coordinating growth and social policies is essential.

In emerging societies and developing countries, it is of crucial importance to reduce dependence on exports and build inclusive development; in the US and UK it is essential to reinvest in inward development. In Brazil social policies including improving education and access to education have become increasingly central to growth strategies (Leahy 2012). In the words of Dilma Rousseff, "Only economic growth, based on income distribution and social inclusion, can generate resources to pay the public debt and cut deficits" (2011). Among emerging societies the policy framework of shared growth is most clearly endorsed in China and Brazil, for political (social stability) as well as economic reasons (domestic demand).

NOTES

1. With thanks for comments to Rudolf Traub-Merz, Manoranjan Mohanty, and Anne Tittor. An earlier version was published in Rudolf Traub-Merz, ed. *Redistribution for Growth?Income Inequality and Economic Recovery.* Friedrich-Ebert-Stiftung: Shanghai Coordination Office for International Cooperation, 2012, 1–11.
2. An increase of the minimum wage (to Bt300), salary increases for civil servants, a reduction of personal income tax (from 30 to 23 percent), a Bt100 million allocation for each province to establish a fund for women, better incomes for the aged, credit cards for farmers, energy cards for taxi drivers, and a tablet PC for all grade 1 students (at the exchange rate of Bt31 for US$1).

REFERENCES

Albert, Michel. 1993. *Capitalism against Capitalism*. London: Whurr.
Bellofiore, R., F. Garibaldo, and J. Halevi. 2011. "The Global Crisis and the Crisis of European Neomercantilism." *Socialist Register* 47: 120–46.
Burchardt, Hans-Jürgen, Anne Tittor, and Nico Weinmann, eds. *Sozialpolitik in globaler Perspektive: Asien, Afrika und Lateinamerica*. Frankfurt: Campus Verlag.
Chenery, H., et al. 1974. *Redistribution with Growth*. Oxford: Oxford University Press.
Commission on Growth and Development. 2010. *The Growth Report: Strategies for Sustained Growth and Inclusive Development*. Washington, DC: World Bank.
Cowen, Tyler. 2011. *The Great Stagnation*. New York: Dutton/Penguin.
Dagdeviren, H., R. van der Hoeven, and J. Weeks. 2002. "Poverty Reduction with Growth and Redistribution." *Development and Change* 33 (3): 383–413.
Etzioni, Amitai. 2011. "Less Is More: The Moral Virtue of Policy Minimalism." *Journal of Global Studies* 2 (1): 15–21.
Fuller, Thomas. 2011. "Empowered, Rural Voters Transform Thai Politics." *New York Times*, July 2.
Hille, K. 2012. "New Generation Feels Weight of the Past." *Financial Times*, November 9, 2.
Im, Hyug Baeg, ed. 2007. *The Social Economy and Social Enterprise*. Seoul, Songjeong Press.

Iverson, T., and A. Wren. 1998. "Equality, Employment and Budgetary Restraint: The Trilemma of the Service Economy." *World Politics*, 50 (4): 507–46.

Leahy, Joe. 2012. Interview with Dilma Rousseff: "We Want a Middle-Class Brazil." *Financial Times*, October 3, 11

Lee, Byeong-Cheon, and Jun Ho Jeong. 2011. "Dynamics of Dualization in Korea: From Developmental Dualization to Exclusive Dualization." Paper presented at Seoul National University Asia Center conference "Global Challenges in Asia."

Li, Peilin. 2012. "China's New Development Stage." In *Globalization and Development in East Asia*, edited by J. Nederveen Pieterse and J. Kim, 141–49. New York: Routledge.

Murphy, Craig. 2006. *The UN Development Programme: A Better Way?* New York: Cambridge University Press.

Myrdal, Gunnar. 1944. *An American Dilemma: The Negro Problem and Modern Democracy.* New York: McGraw-Hill, 1964 reprint.

Nederveen Pieterse, J. 2004. *Globalization or Empire?* New York: Routledge.

———. 2010a. *Development Theory: Deconstructions/Reconstructions.* 2nd ed. London: Sage.

———. 2010b. "Innovate, Innovate! Here Comes American Rebirth." In *Education in the Creative Economy*, edited by Daniel Araya and Michael A. Peters, 401–19. New York: Peter Lang.

Palley, Thomas. 2011. "The End of Export-Led Growth: Implications for Emerging Markets and the Global Economy." Briefing Paper 6. Shanghai: Friedrich Ebert Stiftung Shanghai.

Pogge, Thomas W. 1998. "A Global Resources Dividend." In *Ethics of consumption*, edited by D. Crocker and T. Linden, 501–536. Lanham, MD: Rowman & Littlefield.

Prestowitz, Clyde. 2005. *Three Billion New Capitalists: The Great Shift of Wealth and Power to the East.* New York: Basic Books.

Roach, Stephen S. 2009. *The Next Asia: Opportunities and Challenges for a New Globalization.* Hoboken, NJ: Wiley.

Rousseff, Dilma. 2011. "Brazil Will Fight Back against the Currency Manipulators." *Financial Times*, September 22, 11.

Schmitt, John. 2012. "Economic Development and Income Inequality in the United States since 1979: Inequality as Policy." In *Redistribution for Growth? Income Inequality and Economic Recovery*, edited by Rudolf Traub-Merz, 94–101. Friedrich-Ebert-Stiftung: Shanghai Coordination Office for International Cooperation.

Sumaria, Shaheen. 2010. *Social Insecurity: The Financialisation of Healthcare and Pensions in Developing Countries.* London: Bretton Woods Project.

Vaut, Simon. 2012. "Redistribution for Growth?: Income Inequality and Demand-Led Economic Growth—the German Experience." In *Redistribution for Growth? Income Inequality and Economic Recovery*, edited by Rudolf Traub-Merz, 102–7. Friedrich-Ebert-Stiftung: Shanghai Coordination Office for International Cooperation.

Wen Jiabao. 2011. "How China Plans to Reinforce the Global Recovery." *Financial Times*, June 24, 9.

Wilkinson, Richard, and Kate Pickett. 2009. *The Spirit Level: Why More Equal Societies Almost Always Do Better.* London, Allen Lane

Wodsak, Veronika. 2012. "Reducing Inequality and Promoting Growth through a Social Protection Floor for All." In *Redistribution for Growth? Income Inequality and Economic Recovery*, edited by Rudolf Traub-Merz, 34–41. Friedrich-Ebert-Stiftung: Shanghai Coordination Office for International Cooperation.

4   Brazil's Labor Market
Limitations and Opportunities for Emancipation

Adalberto Cardoso

In the last ten years or so, Brazil has appeared in the world market radar as a credibly emerging economy. High levels of GDP growth, favorable foreign trade leading to high international reserves, job creation to nearly full employment, a politically led and efficient program of income distribution and alleviation of poverty, rising individual and family incomes, public investment in infrastructure and social policies—all that alongside maintaining a strict liberal macroeconomic framework with fiscal austerity, inflation targeting, central bank autonomy, a free and floating exchange rate, and a free financial market as conditions of the entire scenario.

It is not without surprise that 'novelty' is the recurrent word in mainstream economics and also in part of sociological reasoning on Brazil's (uncontestable) recent economic growth and employment creation. In fact, during the entire 1990s the alleged rigidity of the Brazilian labor market has been reputed to be an obstacle to economic restructuring in a globalized world (Pastore 1997; Amadeo and Camargo 1996; Heckman and Pagés 2000) and hence to economic prosperity. Thus like many other Latin American countries that adopted the Washington Consensus recipes, the Fernando H. Cardoso government (1995–2002) introduced a series of flexibility measures to fight unemployment and create jobs.[1] But unemployment more than doubled during his eight-year administration, reaching over 20 percent in some metropolitan regions, and eight out of ten of all jobs created during his term were informal jobs. The Lula administration, on the contrary, cut back some of the flexibility measures, and from 2003 to 2010 the figures were reversed: eight out of ten new jobs were formal ones, totaling 16 million in eight years, against less than 2 million under Cardoso's administration.

While recognizing the deep changes under way and the effectiveness of economic growth and job creation policies, my intention here is to take them *cum grano salis*. In spite of the obvious improvements in the aforementioned dimensions, the country has to face considerable social, economic, and demographic inertia resulting from long-lasting economic stagnation and poor growth rates, which have affected successive generations in the last thirty years or so and have created a new kind of labor market dual dynamic that is seldom taken into account in mainstream analysis and public debate. A good proportion of the 40 percent of the labor force informally occupied in 2009 *are no longer employable* in the formal labor market.[2] That is, they are not demanding and will not demand a formal, salaried job in the growing economy for they are too old and have stayed for too long in either own-account[3] or informal salaried positions, which have selected them out of the competition for the new jobs. In addition, the growing formal labor market *is not* creating enough jobs to accommodate the new generations of workers, who must still face long periods of unemployment or of precarious informal jobs before they are 'entitled' to a formal position. And this position, once attained, is unstable for most workers, representing spans of formal relations interlaced with informality, unemployment, or discouragement. If things are much better today than they used to be for a fair part of the labor force, their stability and sustainability, if delivered in the future, are no guarantee of better labor market positions for what I call here *the lost generations*, which had to make their living in very precarious economic environments in the past, thus burning their bridges to the world of formal work relations, which opens the doors of work-related legal and state protection. This means, on the one hand, that they will increasingly depend on the circulation of the wealth produced elsewhere, of which they will get a share via the (unskilled) services most of them provide and the (low-quality) consumer goods they transform as own-account (or self-employed) workers; and, on the other hand, as they grow older, they will more and more depend on state redistribution and compensatory policies and on family members' support. This social inertia will weigh on the Brazilians' future prospects for emancipation, and must be taken into account in any serious discussion of our future.[4]

In this chapter I will offer a long-run perspective of the characteristics of the Brazilian labor market, in order to scrutinize its present dilemmas and its future prospects. I start with some stylized facts on the economy in the last seventy years, move to an in-depth analysis of changes in the labor market probabilities of men and women of different age groups in the last thirty years, and scrutinize the employment quality of different categories of workers, especially formal sector salaried ones. I also add some remarks on work mobility to demonstrate both the flexibility and the precariousness of the formal labor market, the social foundation of the current socioeconomic 'prosperity'.

PROMISES AND EXPECTATIONS

It is common wisdom that the Brazilian labor market has historically offered precarious jobs, in both urban and rural areas.[5] Precariousness here means long working hours, unhealthy work conditions, despotic work regimes (Burawoy 1979), unskilled jobs, high turnover rates, and low salaries. These resulted from the combined effect, in the last hundred years or so, of highly concentrated land ownership, export-oriented agriculture, low levels of industrialization, high internal rural-urban migration rates, and low investment in education, among other things. The slow emergence of a regulated urban labor market after the 1930s has attracted masses of miserable rural migrants with high levels of illiteracy (80 percent or higher) in search of better living conditions, or of what I call the 'promises of the labor and social rights' (minimum wage, protected jobs, access to social security systems, health and education services), which the new urban areas were simply unable to universalize.

Graph 4.1 presents strong evidence of the adhesion of Brazilian rural-urban migrants to the promises of the urban labor market. It depicts the number of work registry books (*carteiras de trabalho*) issued by the Ministry of Labor from 1930 to 1976,[6] the number of urban formal jobs created, and the increase in the urban Economic Active Population (EAP) during the period. Until 1940 the formal urban labor market had created less than 2 million out of 5 million existing occupations, and the Ministry of Labor had issued less than 800,000 work books. From 1940 to 1950, of the near 1.8 million jobs created, 1.2 million were formal jobs. However, the Ministry of Labor issued 2.7 million work books, equivalent to 230 percent of the number of formal jobs created. This seems to be strong evidence that workers did believe in the *possibility* of being incorporated in the emergent urban formal labor market. This dynamic deepened in the subsequent years. Considering the entire period, the increase in the number of formal jobs was equivalent to only 38 percent of the number of workers that qualified themselves for a formal occupation by getting their work registry books. Reading the evidence differently, it can be said that migrant workers' hopes of inclusion in the formal labor market and economy faced a discount rate of 62 percent, which was the proportion of work book owners that exceeded the number of formal jobs created in urban areas.

That is, the pattern of industrialization based on import substitution has been unable to create enough urban jobs to accommodate the migrant masses. Besides, the jobs created were far from able to match migrants' expectations of better living standards.[7] Many would refuse to work under the offered conditions and salaries, whereas others would get precarious jobs just to 'make a living'.[8] The combination of strong population flows and poor labor market conditions has generated a long-term population inertia characterized by high levels of poverty, underemployment, informality, and social and economic deprivation. Still in 1981, 48 million people, or 40 percent of Brazilians, were below the United Nations' poverty line, which climbed to 43 percent in 1993 (reaching 61 million people), going down to 35 percent in 2001.[9] It is true that the figure was reduced to 21 percent in 2009 (yet comprising 39.6 million people) in the recent process of poverty and inequality alleviation (see Medeiros et al., this volume), but, albeit important in historical terms, it is still not evident whether this reduction will continue in the future due to the current global turbulence and to some structural limits of the Brazilian economy and labor market.

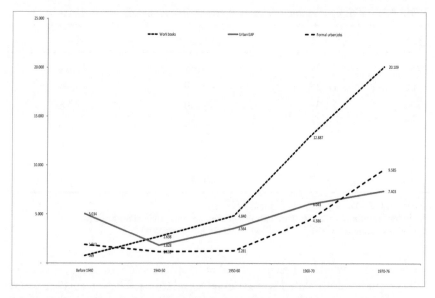

Graph 4.1 Evolution of the urban EAP, of the number of work registry books expedited by the Ministry of Labor, and of the number of formal jobs created (in thousands), Brazil, 1940–76. Source: IBGE, Anuário Estatístico do Brasil, several years; and IPEADATA for estimates of the urban EAP and of contributors to social security. Contribution to social security is taken as a proxy of the creation of formal jobs.

ECONOMY AND LABOR

In 1940 65 percent of the Brazilian Economic Active Population (EAP) were occupied in agriculture, whereas 10 percent worked in manufacturing, construction, and mining. In 1980 the figures were 29 percent and 25 percent respectively. This was the peak share of the three latter industrial groups in employment. In 1990 they occupied 22 percent of the EAP (15 percent in manufacturing), and 19 percent in 2000 (12 percent in manufacturing).[10] Manufacturing alone occupied 13 percent of the workforce in 2009.[11] In terms of employment, then, Brazil has never been an industrial society. It is true that manufacturing, construction, mining (oil included), and urban utilities departed from 25 percent in 1950 to reach a peak of 44 percent share in the gross domestic product (GDP) in 1980 (see Table 4.1). But that share has been declining ever since. The share of manufacturing was a mere 15.8 percent in 2010. In both employment and GDP, manufacturing has occupied a subordinated position in most of the country's recent economic history.

Personal services and small commerce prevailed in the urban job creation dynamic. As a consequence, the Brazilian labor market has always been flexible and precarious over time. That is, the country's labor market

Table 4.1   GDP by Economic Sectors (Percent Share), Brazil, 1950–2010

Year	Agriculture	Industry[a]	Services[b]
1950	25.08	24.96	49.61
1960	18.28	33.19	48.69
1970	12.35	38.30	49.78
1980	10.89	44.09	44.46
1990	8.10	38.69	52.66
2000	5.60	27.73	66.67
2010	5.77	26.82	67.41

Source: IBGE—Department of National Accounts.
[a]Includes manufacturing, construction, mineral extraction, and urban utilities.
[b]Includes commerce. Until 1970, excludes finance intermediation.

is not characterized by 'precarization' or 'flexibilization' of previous universally formal and regulated work and labor relations. These terms are usual in the literature on labor market transitions during the recent neoliberal wave in the OECD countries,[12] meaning a process of deterioration of previously regulated and 'good' jobs. It is true that urbanization in Brazil has meant increasingly better (but still very bad) labor market conditions until the mid-1970s, when the rate of urban formal employment approached 60 percent of the EAP (Cardoso 2003). But this has proven to be a ceiling for the rate of formal labor relations, which fell to near 50 percent and stayed there during the 1980s, and was reduced to near 42 percent in the 1990s just to return to the 50 percent level in recent years.

To be formally or informally occupied is not a minor distinction in Brazil. Formal jobs grant access to a myriad of labor rights that are constitutionalized and encompass both formal standards for collective bargaining, including interest representation and conflict mediation, and substantive rights related to working conditions, health standards, wages, and many other things.[13] As Noronha (2000) put it, the labor law is the most enduring feature of the Brazilian system of labor regulations, the backbone of which dates back to the 1930s. And despite the fact that the costs of noncompliance with the law have historically been either low or null for employers (Cardoso and Lage 2007), the rate of compliance has probably been over 90 percent of the formally occupied labor force in the last two decades.[14]

A formal job is also a 'better' job in terms of income, skilling, and job security. Considering the 1992–2009 period, formal salaried workers earned more than twice as much as informal salaried ones, no matter if men or women. The mean ratio to self-employed workers in the period was 1.66 for men and 1.48 for women in favor of formal wage earners.[15] Formal workers also had two to three more school years than informal ones, and a formal wage earner would stay thirty-six more months (if woman) and twenty-two more (if man) in the same job than an informal one. However, the probability of getting a formal job and staying in it is not the same for all workers, and it competes with a series of other, less 'virtuous' social identities and possibilities.

Looking at the picture with a magnifying glass, Graph 4.2 shows the structure of labor market probabilities of *men* of different age groups, according to the types of jobs or occupational positions available from 1981 to 2009 in Brazil.[16] Each layer in each subgraph depicts the changing positional probabilities, year by year, of a particular age group in the specific job type or social condition. From bottom up, the first layer shows the probability of being in a formal, registered (either public or private sector) job. The subsequent one shows the probability of having an informal, unregistered salaried job in the private sector (by definition, public sector salaried jobs are always formal); the third layer depicts informal own-account jobs; then comes *formal* own-account workers, that is, self-employed men who contribute to a social security system; then come the employers, followed by unpaid workers, unemployed, and men out of the EAP. Reading the data from left to right in each subgraph, the probabilities of a particular age group occupying one of these positions vary over time, and the graphs depict the entire, grouped probabilities for all men of twenty to fifty-nine years of age.

The probabilities for each age group look very stable over time, but some movements must be underlined. In 1981, a twenty- to twenty-four-year-old man had a chance of near 45 percent of being in a formal job. By the mid-1980s this had grown to nearly 50 percent, in what appeared to be a process of labor market structuration and improvement. However, after 1986 the figures fell steadily until a nadir of a meager 34 percent in 1999. The probabilities would then rise again to near 45 percent in 2009, that is, the same figure of the first year in picture (1981). For all age groups the movement is basically the same, although at different probability levels: the twenty-five- to twenty-nine-year-old and thirty- to thirty-four-year-old groups start the period at a 50 percent level, grow to 53 percent, fall to around 42 percent, and rise back to over 50 percent. The older group here depicted starts at a 26 percent level, goes up to 30 percent, down to 22 percent, and back to 27 percent of formal employment probability. This means that, after three decades of economic turbulence (including the 'lost decade' of the 1980s), economic restructuring in the 1990s followed by deep economic crisis and then by the recent growth period, any men of twenty to fifty-nine years in 2009 had basically the same chances of getting a formal job than their age group peers of 1981. And the probability was almost always below 40 percent except for a man of thirty to thirty-four years.

Because we're analyzing different generations entering the labor market in different points in time, what is important here is that probabilities in a given point are not neutral with respect to future labor market prospects of each age group. We know from literature that an unemployment spell has further consequences for a young person's career; the duration of unemployment is also important, and so is the type and quality of the first job attained. Bad labor market conditions resulting from economic crisis create period effects affecting all workers in a given historical juncture, but with different consequences over time for different age cohorts, a person's skills, sex, and other intervening factors.[17]

For instance, a good proportion of skilled workers that lost their jobs in the metallurgic belt of the São Paulo Metropolitan Region during the 1981–83 recession never returned to a formal job again (Hirata and Humphrey 1989). We also know that the economic restructuring of the 1990s destroyed near 1.4 million formal jobs in manufacturing in Brazil (Sabóia 2000). These jobs were not recovered before the mid-2000s, and the manufacturing workers that had lost their jobs were now too old to be 'employable'. As a matter of fact, considering auto-industry workers dismissed after 1989, less than 50 percent would ever find another formal job again (Cardoso, Comin, and Guimarães 2004). Besides, the older the worker the less likely it is for him or her to get another formal job. So we must always consider cohort and period effects jointly in the analysis of labor market probabilities and impacts on workers' life cycles. I will return to this later.

Another general and important movement of the global probabilities of men is the fact that *informal salaried* positions are constantly reduced over the life cycle in favor both of formal salaried and self-employment, no matter the period (or year). Informal salaried relations are important entry conditions for younger men, and lose importance as they grow older. Probabilities were around 22 percent or more for the youngest age group, and 12 percent or less for the oldest group here depicted, no matter the period in the 1981–2009 time span. On the other hand, for each age group the probabilities of an informal salaried job are basically constant over time. In other words, people aged thirty to thirty-four years in 2009 had the same probability of their 1999, 1989, or 1981 peers of being informally occupied, varying very little around the mean of 16.4 percent (standard deviation of less than 1 percent). The proportion is basically the same for the forty age group.

The combination of these two probability movements (reduction with age and stability over the thirty-year period for any age group) suggests that informal salaried relations are transitory conditions for a fair proportion of the *younger* workers, who take them while awaiting a better, formal position. This seems to be evidence that this structure of job positions actually offers opportunities for job and social mobility, because informal salaried jobs are traditionally worse paid than the other two categories. For a portion of the younger men entering the labor market, then, it presents itself as a structure of opportunities of mobility that is actually instantiated during a person's life cycle.

But the sequence of graphs offers other crucial evidence. For all age groups, men's probabilities of being either unemployed or out of the EAP (the upper two layers) increased importantly in the last thirty years. For the youngest age group of men, if we include the probability of being in an

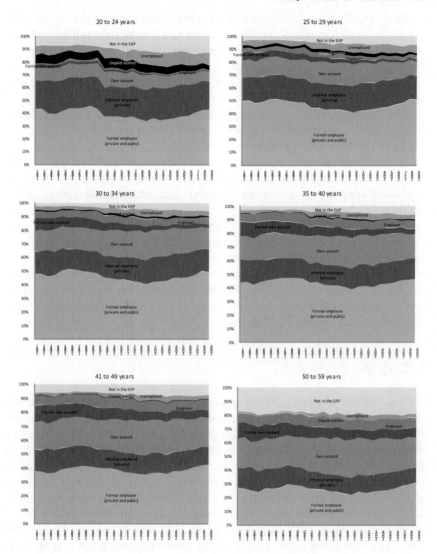

Graph 4.2 Type of job by age groups: men of twenty to fifty-nine years, Brazil, 1981–2009.

unpaid job, after a nadir of 17 percent in 1990, the joint probability of these three positions grew to 30 percent in 2003, when it started to fall again, but only to reach 26 percent in 2009. If this can be taken as a measure of bad labor market or life cycle positions,[18] all age groups were worse off in this particular matter in 2009, compared to 1981. The least affected were thirty- to forty-year-old men, but even for them these three positions' probabilities increased from 5 to 10 percent in 30 years.

Note also that, as men grow older (that is, comparing the age groups vertically in the graphs' set), there is a clear exchange between formal and informal salaried probabilities, on the one hand, and own-accounting, on the other. For each older age group, the proportion of the first two positions shrink while own-accounting grows, and that includes both formal and informal own-account jobs.

The message so far seems straightforward. The peak 'virtuous' labor market age for men in Brazil is from twenty-five to thirty-four years. Here we have the maximum participation in the EAP, the minimum unemployment rate, and the higher proportion of formal positions (if we include formal salaried and own-account jobs, and employer positions). Younger and older age groups have relatively higher (sometimes much higher) probabilities of being either out of any job or in an informal job. In any case, life cycles are marked by a structure of occupational opportunities that is very age specific and that worsens importantly with age. Aging has not been synonym of work-based ontological security so far for a very large proportion of the EAP. And most importantly, recent labor market improvements have not decisively changed this fate. Before elaborating on the consequence of this for the current debates on economic recovery and social change, I must scrutinize the quality of formal jobs.

THE QUALITY OF JOBS

One important indicator of the quality of labor market conditions is the degree of stability of the existing positions. If bad jobs (in terms of income and labor rights) are attained for large periods of time, they must be taken as the very boundaries of the social classification possibilities of their occupants. For instance, informal salaried relations' probabilities are reduced as people grow older, as we've seen, but those who are found in these positions at the age of, say, forty-five years, have most probably been there for quite a long time, and the precarious positions not only will have scratched a scar on their past lives, but will also weigh on the workers' future. To analyze this I propose Graph 4.3, which shows two complementary scenarios of men's labor market positions. On top, we see the mean duration of employment in months, of all men of fifteen years or older occupied in Brazil from 1992 to 2009,[19] and at the bottom, the employment duration of those of forty to forty-nine years of age. We can see, first, that employment duration of all categories is increasing over time, though at different paces. Own-account workers (formal and informal) and employers stay much longer in their positions than the other three categories, and duration increased significantly throughout the period. In 1992 their mean job duration varied from 135 (formal own-account) to 150 months (employers). In 2009 the duration had increased to 150 (informal own-account) to 170 months (formal ones). For the other three categories there has been a slighter increase, and at lower duration levels: from 75 to 85 months in the case of formal employees and unpaid workers, and below 60 for informal salaried jobs. This is evidence of the transitory character of the latter for a fair part of the labor force. It also means that own-account jobs are a repository of labor to be mobilized by an expansion of labor demand due to economic growth, but a repository with clear oversupply.

In fact, the second part of the graph shows an important increase in the job duration of own-account workers of the forty to forty-nine age group, from less than 150 in 1992 to almost 170 months in 2009. Comparing this figure with Graph 4.2 above, in which a proportion of own-account prob- abilities were transferred to formal employment after 2003, the increase in own-account duration results from a *selection* of longer-lasting jobs in this category. In other words, those who moved from own-account to formal employment were people that were relatively neophyte in the former posi- tion, probably waiting for a salaried formal job in better junctures. This is expressed, in reversed mode, in the slight reduction in the mean duration of formal jobs after 2003 (from 120 to 116 months) in this particular age group, reflecting the entry of new occupants in the jobs then created.[20] In the same direction, older workers remain longer in informal *salaried* posi- tions than the others. Here too a selection process after 2003 seems to have happened, for job duration increased from the 'historical' figure of 80 to almost 110 months. This is why I said that if a forty-five-year-old man is found in this position in 2009, he has probably been there for quite a long time, and was selected out of the probability of entering a formal position.

Taken together, these figures seem to be telling the following story. Workers tend to stay a shorter time in 'good', formal jobs, especially in private ones, and a longer time in informal, own-account positions. The economic recovery after 2003 has generated 11.5 million new formal jobs until 2009 (including public and private sectors).[21] These were distributed mostly among people already in the labor market, employed in either a formal or an informal existing position. New entrants also got a share of it, but the increase in the unemployment rates of younger age groups (seen in Graph 4.2) suggests that they benefited much less than those with longer labor market experience.[22] This also tells a story of job mobility and labor market fluidity (or flexibility), because the joint probabilities of different positions are changing for all age categories at the same time.

These figures call attention to an important characteristic of the recent changes in the Brazilian labor market. Job mobility from informal to for- mal positions is apparently restricted *to the fringes* of the two major infor- mal categories (unregistered salaried and own-account jobs). These fringes were composed of workers with less labor market experience and lower salaries than those that stayed, or were selected out of the possibility of getting a formal job. This contributed to increase the median salaries and the job duration of those who survived in the worse positions, while reduc- ing the duration of the new formal jobs, which were occupied in part by

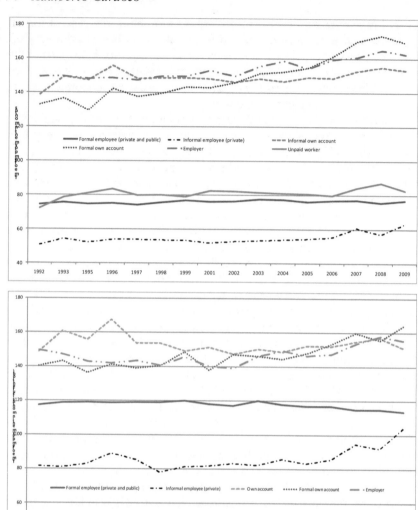

Graph 4.3   Mean time of employment (months) in the current job, by type of job, Brazil: 1992–2009. 1. Men of fifteen years or older. 2. Men of forty to forty-nine years. Source: PNAD.

previously informal workers in search for better salaries and benefits in formal positions. But all this happened to a very small fraction of the labor force, although fairly well distributed within the age categories. Only 10 to 15 percent of the joint distribution of job type probabilities for men moved from one category to another in recent years, with formal jobs in 2009 returning to the proportion they had in 1981. This has not been enough to change the enduring and entrenched structure of probabilities of getting a job, which is still characterized by precariousness and insecurity. For instance, according to the same 2009 PNAD, four occupational groups accounted for 75 percent of the forty- to fifty-nine-year-old informal own-account jobs of men: agriculture (30 percent), construction (25 percent), sales (15 percent), and driving (6 percent). Formal own-account in the same groups were 63 percent. Sales occupied one-third of informal own-account women, and 30 percent of formal ones. Most importantly, a larger proportion of this aging labor force is staying longer in these worse positions, which cannot be treated as a passage point for workers waiting for a better fate. They are the destiny of the large majority of them.

A NEW MIDDLE CLASS?

These findings pose important structural limits to the arguments celebrating the emergence of a new middle class in Brazil[23] and to any inclusive project of emancipation. Self-employment and informal salaried positions are no synonyms of freedom or autonomy from an employer, nor of unconstrained working conditions. In 2009, 20 percent of informal own-account workers worked fifty hours a week or more (compared to 11 percent of formal wage earners). Besides, men would have stayed for 153 months in this position in 2009, and their mean school years were only 5.6. For many, self-employment is a residual situation, not a position previously desired or purposively manufactured during the life cycle. By the same token, a good part of the informal salaried workers earn enough to qualify as part of the emergent 'middle class', but their grip in the economic structure is insecure and unstable, their jobs are lost at any signal of economic crisis, and their next job will not necessarily match the previous standards. Most of all, considering the 1992–2009 period, 64 percent of them figured among the 40 percent lower stratum of the mean income's distribution (informal own-account rate was 52 percent, compared to 26 percent of formal wage earners).[24]

If the economy does not offer sufficient conditions for stable and protected work, the mechanisms that produce and reproduce social and economic inequalities will keep on operating fully fledged. Besides, income is increasing for most of the occupied labor force, as Neri (2010) and others rightly argue, but this is happening in a labor market structure that is precarious and very unstable for young and also for mature workers of thirty-five years and older. Looking at the picture from a life cycle perspective, the challenge of the growing economy is not only to create good jobs for the new generations, but to assure quality social positions and income to the older ones, all this in a very unstable and precarious environment.

Because most self-employed and unregistered salaried workers are located in small commerce and services to persons and families, their income is directly dependent on the wealth produced elsewhere and appropriated by wage earners of all sorts, who will spend their money with goods and services informal workers deliver. The increase in the mass of salaries will certainly benefit them, and this may happen in a way that some of the self-employed workers may (and most probably will) become small employers and even formalize their business, thus registering their informal workers, paying taxes, demanding credit—in a word, helping move the economy while being moved by it. Economic growth may bring about the amelioration of the economic and social conditions of informal workers in a way that may actually help them depart from the realm of necessity. But the country is still far from this scenario.

YOUNGER AND NEW GENERATIONS

In 1992, when the PNAD collected the information for the first time, 36 percent of men and 30 percent of women declared to have started working at the age of ten years or less.[25] Those that began at fifteen years or less were 80 percent of men and 68 percent of women. This is one of the various measures of the *cost of the past* that framed the working population profile at that juncture. And in fact, Ribeiro (2007) has shown that, in 1996, some 80 percent of the Brazilians, considering parents and their offspring, had experienced a rural-urban migration, and in rural areas child work was a rule. The inertia of these figures is astounding: seventeen years later (2009), we still find 25 percent of men and 18 percent of women declaring to have started working at the age of ten or less. Two-thirds of men had started at the age of fifteen or less. If we consider twenty-five-year-old men only, 58 percent had started working before completing sixteen years in 2009 (compared to 44 percent of young women). This is proof that the beginning of working life is being postponed throughout time, but at a very low pace. Even the new generations start to work very early.

Most importantly, at the age of eighteen, 50 percent of males and 46 percent of females were already out of school in 2009 (compared to 62 percent and 56 percent in 1992), and these figures were not equally distributed among the social classes. In families that earned three minimum salaries per capita or more (near US$700 per capita per month) in 2009, 80 percent or more of their eighteen-year-old boys were still at school, whereas family income of half minimum wage or less was associated with a 40 to 50 percent probability of their eighteen-year-old adolescents attending schools.

Early school leaving has long-lasting consequences for a person's working life, but even today the school/work trade-off still favors work, especially in poorer families. One-half to two-thirds of the wealthier families (those that earn two minimum wages per capita or more) happen to take their siblings to higher education, but the figure is no higher than 15 percent for families earning one minimum wage per capita or less. Many men and women will return to school later, trying to catch up with labor market changes and the new demands for skilled workers in the emerging sectors.[26]

It should be noted that choosing work against school was a *rational* decision of poorer families in the stagnated Brazil of the 1980s and 1990s. Access to quality universities was (and still is) a function of strong investment in early, private education, something difficult to achieve in poor households. And even if families decided to choose some kind of deprivation to favor part of their children, the prospects of a good job after schooling were never clear, due to the ups and downs of labor market opportunities in more than twenty years of meager economic growth. The new generations are still paying the price of their parents' decisions, and many are making decisions as if the country were still sailing in the mist, with no clear future ahead.[27]

CONCLUSION

The informality of work and employment relations is still very high in Brazil. Most men and women start their working lives in an informal occupation, at an early age, most often before completing eleven years of schooling. Informal salaried and own-account positions are entrance doors and also a repository of labor for the formal labor market. When the latter expands rapidly, like in the last eight years, it feeds from the former mechanisms of making a living. This means that, looking at the picture from the standpoint of the occupied labor force, the labor market is made of the entire set of those mechanisms, which are mobilized in different moments of the workers' biographies according to a logic that combines individual (and family) strategies with the actual structure of opportunities, a logic the element of which is, most often (but not always), the need to survive.

The problem, from an emancipation point of view, is that for a large proportion of the working population, the informal 'segment' of the labor market is not a transitory terrain, nor a reception hall for those waiting for better positions. Rather, it is the *dead end* of their work histories. The country will not create good and protected jobs for the majority of these workers, either because they are not sufficiently skilled; or because of age discrimination in a labor market with oversupply, which allows firms to opt for younger workers willing to exchange school time for increasingly better-paid jobs; or because of the dilapidation of the bodies of men and women by the precarious work conditions they had to face in their working histories during stagnation; or because of the pattern of development the country seems to be pursuing, based on commodity exports with minor counterpart of high-quality manufacturing, service, and commerce sectors, the consequence of which is the continuous creation of bad-quality jobs, etc.[28]

In this scenario, the country has to face what I have called the 'cost of the past', a concept that denotes a set of economic and demographic dynamics that has punished generations of workers with low growth and precarious jobs, meaning unprotected and badly paid. The challenge is to find means to alleviate the future deprivation of these workers and, at the same time, create good jobs for the new generations. Economic development along with productive inclusion may fulfill this task in the medium run. Young people of twenty-nine years or under have appropriated some 30 percent of the new jobs created in recent years, and their unemployment rate is among the lowest in the world (around 11 percent in 2010). However, their informality rates are still very high, and their salaries are still consistently lower than that of the older generations, despite their higher levels of education. Besides, 70 percent of the new jobs for men were occupied by people thirty years of age or older, with previous labor market experience, most often in informal positions. However, this exchange comprised a minor proportion of the occupied labor force, so that formality rate could return to only its 1981 level. It is true that formal salaried relations are expanding fast in Brazil, but the social debt accumulated in decades of precariousness is still very high.

In that respect, the response to the first challenge (safeguard the older generations) will fall on the shoulders of the state's redistributive capacity, and on its ability to recognize these 'lost generations' the right to a dignified life, in an environment inhospitable to the instantiation of this right. This environment is still molded by precarious and unprotected mechanisms of making a living, and it is also framed by important macroeconomic restrictions to redistributive policies.

Good-quality labor markets are crucial for emancipation. In a capitalist society the labor market is still the central way of access to social positions; to conditions that give way and sustain individual and collective identities; to income; and to an important social right: a decent retirement. Labor markets respond to investment: public (demanding services and infrastructure, expanding public services, etc.), private (especially in the case of labor intensive sectors), and families (in their members' skills and quality of life). In the ten years that ended in 2010, the rate of investment in Brazil has varied little around 17 percent of GDP (as against 40 percent in China and 22 percent in many OECD countries), and the increase in workers' schooling has been very slow by any international standard.[29] Better and more stable labor market conditions require higher rates of investment of all economic agents, and government of the three administration levels (federal, provincial, and municipal) should face the challenge of leading the economy's investment drive and of offering much better education services.

It is true that the rate of investment in education is increasing in recent years. After decades of underinvestment, in 2009 public expenditure per pupil in primary education was equivalent to 20 percent of the GDP per capita, just a bit below the US and the high-income OECD countries.[30] The problem, of course, is that GDP per capita was under US$10,000 in Brazil and over US$30,000 in the two mentioned areas. The equivalent proportions hide the relative poverty of the available resources in Brazil. So the country must assure economic growth but also increase the proportion of GDP invested in primary and, most especially, secondary education.[31] This is urgent, although the labor market consequences appear only in the medium and long run. It must also stimulate labor-intensive sectors in services and manufacturing.

Labor markets also respond to institutional settings and labor regulations. Most agents would agree that the labor law is hard to abide by micro- and small business. The labor movement and the left-wing Workers' Party (in office since 2003) have always resisted any attempt to reform the labor code for fear of deregulation and flexibility. But a concerted (between capital, labor, and state), state-led project of reduction of the labor costs could increase the formality rate, labor productivity, and even real wages in small business, and stimulate better working conditions for men and women.

NOTES

1. Among the flexibility measures, we can name the legal institution of flexible working hours, incentives to increase the variable proportion of wages (via institution of workers' share in profits, productivity bonuses, fringe benefits, etc.), reduction of dismissal costs, incentives to collective bargaining instead of the protection of the labor law, and many others. See Cook (2007) for a Latin American perspective on liberal labor market flexibilization. A good critical assessment of the trade-off between flexibility and employment creation is Bensusan (2006).
2. On similar lines, see Baltar et al. (2010).
3. In Brazil own-account work means work in one's own business, exploring an economic activity without employees, either individually or with a partner, with or without the aid of nonsalaried worker(s). See http://www.ibge.gov.br/home/estatistica/populacao/condicaodevida/ indicadoresminimos/ conceitos.shtm. Own-accounting is traditionally associated with low income and informality, although a recent increase in entrepreneurship in the country is creating new own-account strata at the top of the income's distribution. I will return to the point in this chapter.
4. The same problem can be found in India, although probably less in China. In India life expectancy is increasing fast and is about to overtake that of Western rich countries (ILO 2011). Most of the older workers have not contributed to social security systems, for they've lived in the informality for their entire lives. In Brazil, as will be seen, the experience of a formal job was fulfilled here and then for many, but never universalized as a stable condition.
5. See Oliveira (1972) and Cardoso (2010).
6. The work registry book (*carteira de trabalho*) was created in the early 1930s. Every formal job generates a record in the book with information on salary, working hours, occupation, annual paid leave, and a few other items. It also registers the end of the job, including the reason (just cause, voluntary leave, retirement, etc.). If a job is registered in the worker's book, it is because the employer is (at least formally) willing to comply with the labor law.
7. Important studies on the frustration of rural migrants in emerging urban labor markets are in Lopes (1967). The author attributes the frustration to the maladjustment of rural migrants to manufacturing work routines. Touraine (1961) maintains the idea of a "Utopian nonconformity" of these migrants, who refused precarious and low-paying jobs that did not match their upward mobility expectations.

8. A good and still compelling argument on the matter is Machado da Silva (1971).
9. Souce: IPEADATA (http://www.ipeadata.gov.br).
10. Data from decennial censuses, in IBGE (2003).
11. Data from the National Household Sample Survey (PNAD).
12. For instance, Castel (1995), Standing (1999, 2012), and Sennet (2003).
13. The regulations firstly constitutionalized in 1934 and extended in the Constitutions of 1937, 1946, and 1988 include: working hours; the prohibition of night work for women and youth; minimum working age; entitlement to one day off each week; special rights for women during and after pregnancy; the definition of a minimum wage based on the basic needs of a worker's family; equal pay for equal work; salary protection; limits on overtime; the right to housing and schooling; employer responsibility for work-related accidents and diseases; minimum occupational safety and health standards; the right of association for workers and employers; the right to strike; tripartite bodies for conflict resolution; labor courts; compensation for unjust dismissal; and the no renounceable character of labor rights.
14. The number of workers' demands at the first level of the labor courts varies from 2 to 2.5 million every year, for a formal occupied population varying from 25 to 42 million in the last twenty years.
15. Data from National Household Sample Survey (PNAD). The differences are lessening. In 1992 a formal employee earned almost three times as much as an informal one, a ratio reduced to 1.8 in 2009, still an important difference. Formal wage earner to own account ratio was reduced from 1.94 to 1.57 (men) and from 1.78 to 1.39 (women). In any event, all ratios have been fairly stable over the last five years.
16. The source is the National Household Sample Survey (PNAD). Due to space limits I will analyze only men's probabilities in depth. The picture for women is very different and will be discussed in passing later.
17. Literature on the matter abounds. A good study for the case of England is Purcell, Wilton, and Elias (2007). For Brazil, Guimarães, Silva, and Farbelow (2008).
18. It is true that part of the young men out of the EAP were actually studying. So, for them being out of the labor market is not necessarily a 'bad' condition as compared to older men in the same situation.
19. The question on time in the current job was introduced in the National Household Survey in 1992 only.
20. I've selected this age group for the clear exchange of informal to formal positions after 2003, clearly reflected in changes in job durations. But the movement was also detected in the other age groups, most especially from twenty-five to thirty-nine years.
21. According to the National Household Survey analyzed here, there existed 29.5 million formal salaried positions in the countries' labor market in 2002. In 2009 this had grown to 41 million.
22. As a matter of fact, 70 percent of the new formal jobs created in the period were occupied by people thirty years of age or more, according to the same PNAD.
23. The most important proposition of this theses can be found in Neri (2010). The 'new middle class' is now a research program at Getulio Vargas Foundation, coordinated by Neri.
24. All data computed directly from PNAD 1992–2009.
25. All figures in this section are computed from PNAD microdata.
26. In recent years manufacturers and employers of all sorts complain that Brazil is in face of a shortage of skilled workers, which is presumably limiting the

potential for economic growth. But most experts in the field contend that this is not the case. See Barbosa Filho, Pessôa, and Veloso (2010) and Maciante and Araújo (2011).

27. According to the newly released 2010 Census data, only 25 percent of the twenty- to twenty-four-year-old Brazilians were studying. See http://www. censo2010.ibge.gov.br/resultados_do_censo2010.php, accessed on April 27, 2012.
28. In 1997 the economist Luciano Coutinho proposed the term 'regressive specialization' to characterize the economic restructuring in the 1990s. The concept referred to a process in which large-scale producers of standardized goods, intensive in energy and natural resources and with low value-added capacity, are favored by the overall economic environment, and expand more rapidly than the other sectors. David Kupfer, an expert in manufacturing restructuring, argues that the process is deepening in the 2000s. See Coutinho (1997) and Kupfer (2004).
29. Compared to Korea in the 1970s and 1980s, and to China and India in recent years, as well as to Chile, Argentina, and Uruguay, Brazil stands far behind in mean school years of its population and also in the quality of education, measured by international comparative math and language exams.
30. In 1999 the rate was 10 percent, and much less than that in the 1980s. For the recent period, see World Bank data at http://data.worldbank.org/indicator/SE.XPD.PRIM.PC.ZS.
31. Investment per pupil in secondary education was 20 percent of GDP per capita in 2009, 25 percent less than the expenditure of OECD countries. See World Bank data at http://data.worldbank.org/indicator/SE.XPD.SECO.PC.ZS.

REFERENCES

Amadeo, Edward, and José M. Camargo. 1996. "Instituições e mercado de trabalho no Brasil." In *Flexibilidade do mercado de trabalho no Brasil*, edited by José M. Camargo, 35–57. Rio de Janeiro: FGV.
Baltar, Paulo A., Eugenia T. Leone, MAlexandre G. Maia, Salas, Carlos Salas, José Dari Krein, Amilton Moretto, Marcelo W. Proni, and Anselmo Santos. 2010. "Moving towards Decent Work. Labour in the Lula Government: Reflections on Recent Brazilian Experience." Global Labour University Working Papers, 9, 1–38. http://www.global-labour-university.org/fileadmin/GLU_Working_Papers/GLU_WP_No.9.pdf. (accessed September 2011).
Barbosa Filho, Fernando H., Samuel A. Pessôa, and Fernando A. Veloso. 2010. "Evolução da Produtividade Total dos Fatores na Economia Brasileira com Ênfase no Capital Humano—1992–2007." *Revista Brasileira de Economia* 64 (2): 91–113–000.
Bensusán, Graciela, coord. 2006. *Instituições Trabalhistas na América Latina: Desenho Legal e Desempenho Real*. Rio de Janeiro: Revan.
Burawoy, Michael. 1979. *Manufacturing consent: Changes in the labor process under monopoly capitalism*. Chicago: University of Chicago Press.
———. 2003. *A Década Neoliberal e a Crise dos Sindicatos no Brasil*. São Paulo: Boitempo.
———. 2010. *A construção da sociedade do trabalho no Brasil*. Rio de Janeiro: FGV.
Cardoso, Adalberto, Alvaro Augusto Comin, and Nadya A. Guimarães. 2004. "Le Rejetés de la modernisation." *Sociologie du Travail* 49: 54–68.

Cardoso, Adalberto, and Telma Lage. 2007. *As normas e os fatos: desenho e efetividade das instituições de regulação do mercado de trabalho no Brasil*. Rio de Janeiro: FGV.

Castel, Robert. 1995. *Les métamorphoses de la question sociale*. Paris: Fayard.

Cook, Maria L. 2007. *The Politics of Labor Reform in Latin America: Between Flexibility and Rights*. University Park, PA: Penn State University Press.

Coutinho, Luciano. 1997. "A especialização regressiva: um balanço do desempenho industrial pós-estabilização ." In *Brasil: desafios de um país em transformação*, edited by Reis Velloso. Rio de Janeiro: Fórum Nacional; José Olympio Editora.

Guimarães, Nadya, Paulo H. Silva, and Marcus V. Farbelow. 2008. "A experiência desigual do desemprego recorrente: Diferenças de gênero e raça nas transições ocupacionais em São Paulo." In *Mercados de trabalho e oportunidades*, edited by Nadya Guimarães et al., 118–139. Rio de Janeiro: FGV/FAPERJ.

Heckman, James, and Carmen Pagés. 2000. "The Cost of Job Security Regulation: Evidence from Latin American Labor Markets." Working Paper No. 430. Washington, DC: Inter-American Development Bank.

Hirata, Helena, and John Humphrey. 1989. "Trabalhadores desempregados: trajetórias de operárias e operários industriais no Brasil." *Revista Brasileira de Ciências Sociais* 11 (5): 000–000.

IBGE. 2003. *Estatísticas do Século XX*. Rio de Janeiro: IBGE.

ILO. 2011. "World Parliament of Labour Turns 100." *World of Work* 71 (April). http://www.ilo.org/global/publications/magazines-and-journals/world-of-work-magazine/issues/WCMS_154579/lang—en/index.htm (accessed January 2012).

Kupfer, David. 2004. "Política Industrial." *Econômica* 5 (2): 281–98.

Lopes, Juarez B. 1967. *Crise do Brasil arcaico*. São Paulo: Difusão Europeia do Livro.

Machado da Silva, Luiz Antonio. 1971. "Mercados metropolitanos de trabalho manual e marginalidade." Master's thesis, Museu Nacional—UFRJ.

Maciante, Aguinaldo N., and Thiago C. Araújo. 2011) "Requerimento técnico por engenheiros no Brasil até 2020." *Radar: tecnologia, produção e comércio exterior* 12: 43–54.

Neri, Marcelo C. 2010. *A nova classe média: o lado brilhante dos pobres* [The new middle class: The bright side of the poor]. Rio de Janeiro: FGV/CPS.

Noronha, Eduardo G. (2000), *Entre a Lei e a Arbitrariedade: mercados e relações de trabalho no Brasil*. São Paulo: LTr.

Oliveira, Francisco. 1972. *Economia brasileira: crítica da razão dualista*. São Paulo/Petrópolis: Cebrap/Vozes.

Pastore, José. 1997. *A agonia do emprego*. São Paulo: LTr.

Purcell, Kate, N. Wilton, and Peter Elias. 2007. "Hard Lessons for Lifelong Learners? Mature Graduates and Mass Higher Education." *Higher Education Quarterly* 61 (1): 57–82.

Ribeiro, Carlos Costa. 2007. *Estrutura de Classe e Mobilidade Social no Brasil*. Bauru: EDUSC.

Sabóia João. 2000. Desconcentração industrial no Brasil nos anos 90: um enfoque regional. *Pesquisa e Planejamento Econômico* 30 (1): 69–116.

Sennet, Richard. 2003. *Respect in a World of Inequality*. London: Norton.

Standing, Guy. 1999. *Global Labour Flexibility: Seeking Distributive Justice*. London: Blackwell.

———. 2012. *The Precariat: The New Dangerous Class*. London: Bloomsbury.

Touraine, Alain. 1961. Industrialization et conscience ouvrière à São Paulo. *Sociologie du Travail*, 3 (4): 77–95, oct-dec.

5 Social Movements and Emancipation in Brazil

Ilse Scherer-Warren

INTRODUCTION

Brazilian movements and civic organizations have been participating in transnational mobilizations for human rights, ecology, peace, social justice, alter-globalization, and other similar aims. These interactions promote and consolidate universal humanitarian values arising in the contemporary world in Brazilian social movements, and also promote recognition of the singularity of the demands for changes coming from populations that have been historically excluded, such as the landless, the indigenous, black people, garbage collectors, and so on.

A different question is what do global protest movements tell us nowadays—in 2010 and after people in several Arab countries took to the streets to protest against dictatorial governments, in the Arab Spring?[1] Some analysts concluded that these manifestations were the result of social networking via the Internet, which is partially true because this new form of connectivity is very common and relevant to the networking in contemporary protest movements. However, we can also observe that these movements are gaining broader public recognition because of mass manifestations in the public sphere, especially when the movement has a popular local basis. Physical and face-to-face networks may be the most important elements for empowerment of a protest movement, as demonstrated by the example in Brazil of MST, the Social Movement of the Landless, our movement with the widest transnational visibility. MST recognizes the importance of the Internet as an alternative, more democratic form of media and advocates it as a way to construct a worldwide network for social justice; but at the same time MST recognizes that the most relevant form of empowerment comes from its capacity to organize and mobilize the landless in their local communities and through protest marches.

Analysts of the Arab protest movements observe that they struggle for 'bread', that is, for quality of life, equality, better distribution of wealth and resources, etc., in other words, for socioeconomic democratization; for 'social justice', with reference to human rights in a broader sense,[2] and equity, i.e. for social-cultural democratization; and for 'dignity and citizenship', calling for

Table 5.1   Forums and Networks of Research (Project AMFES)

Forums and Networks of Research (Project AMFES)[4]	Year of Creation
ABONG—Association of Brazilian NGOs	1991
AMB—Articulation of Brazilian Women	1994
FNPETI—National Forum for Prevention and Eradication of Child Labor	1994
FNRA—National Forum for Agrarian Reform and Justice in the Countryside	1995
FENDH—National Forum of Entities for Human Rights	1996
FLC—National Forum about Garbage and Citizenship	1998
FNMN—National Forum for Black Women	2001
INTER-REDES—Network for Rights and Politics	2002
FBO—Brazilian Budget Forum	2002
FBES—Brazilian Forum of the Solidarity Economy	2003
FDDI—Forum for Defense of Indigenous Rights	2004
FSM or WSF—World Social Forum[5]	2001

Source: Scherer-Warren (2012).

political recognition, autonomy for public manifestations, and democratic participation in the local and national public sphere, in other words, they are working towards political democratization.

These ideals and demands are also present in European, North American, and Latin American countries just as they are in Brazilian movements. They express the globalization of symbolic representations of social movement demands and also express the singularity of historical struggles of organized civil society in each country. Thus, to go more deeply into the singularity of each national case we need to research specific social movements and civil organizations, considering the way they articulate their discourses politically for the creation of public agendas with common grounds. Therefore I will present some results of my investigation into forums and networks of civil society and social movements in Brazil, to bring to light some of these discursive articulations[3] in the practices of the following entities listed in Table 5.1.

WHAT IS RELEVANT TO AN UNDERSTANDING OF BRAZILIAN MOVEMENTS NOWADAYS?

I propose three analytical dimensions to characterize contemporary Brazilian social movements that are intrinsically related in contemporary Brazilian social movement practices and that allow us to observe how they may focus positively on the public policies: (1) the organizational frame: networking from civil society to social movement; (2) substantive issues and ideals: material, symbolic and empowering; and (3) advocacy and participation in public policies: the institutional relations.

The Organizational Frame: Networking from Civil Society to Social Movement

In research on civil organizations in Brazil, we observed that as the new technologies of information and communication have had impact and created new forms of social networks, they have also promoted new possibilities of interactions for social movement formations, and as a result we arrive at the organizational frame[6] outlined in Table 5.2, with diverse levels of civil society organization participating in the rise of a social movement network.

To the extent that social movement networks incorporate the organizational, articulatory, and mobilizing levels, as represented in Table 5.2, they transcend the merely empirical expressions of these levels towards the formation of a movement's logic that promotes the construction of a social, political, cultural, or ideological identity, the definition of conflicts, antagonisms, or social and systemic adversaries, and a struggle in the name of a project or

Table 5.2   Format of Organized Civil Society for Networking

Localized Organizational Level
Specific social entities and localized movement organizations
(Such as civil associations, trade unions, settlements, CEBs, church pastorals, NGOs, etc.)
Articulatory Political Level
Civil society forums and interorganizational networks
(such as thematic forums, popular assemblies, organizational networks of social entities)
Mobilizing Level in the Public Sphere
Protests and collective manifestations, campaigns, 'weeks', marches
(Such as the World March of Women, Land Reform March, the Cry of the Excluded, etc.)
Social Movement Network
The set of practices and politics formed by the three levels above
(through interaction based on common identity, conflict definition, and ideals for change)

utopia of social, cultural, political, or systemic change.[7] In other words, they search for some kind of social, political, or cultural emancipation.

Regarding the struggles against social exclusion and for human rights and citizenship, the social forums and organizational networks of our research indicated several social movements as being particularly relevant to the networking of political processes, but they all highlighted the MST. Around their specific demands, the following movements or organizations were indicated by at least two forums: feminism, human rights, the solidarity economy, the unemployed, garbage and citizenship, the budget, eradication of child labor, street children, the indigenous, black people, hip hop, gays, trade unions, the environment, the homeless, family farming, and 'Via Campesina'. Several other demands had only one indication.

Among these movements MST was identified as a strong link in the social networking mobilizations in Brazil and globally. MST's relevance was due not only to its participation as a member of several of these forums but mainly to its capacity of leadership (strategic link) in the broader organization of social movement networks in which the forums are reported. It was also noted for its visibility in the public mobilizing networks and for the political continuity and coherence of its daily activities on the local organizational level.

Substantive Issues and Ideals: Material, Symbolic, and Empowering

We could start by asking "What moves the civil society and how do movements arise or what are their demands, protests and ideals?" In our research we found three complementary dimensions or substantive issues:

- The existential material conditions: which include struggles against inequality, poverty, unemployment, territorial segregation, and the defense of life quality; in other words, this dimension refers to the emblem of *'bread and territory'* and, therefore, it refers to an encounter for socioeconomic democratization, mentioned before. And it is related to the second dimension:
- The symbolic conditions in the social reproduction: which can be represented by the struggles against stigma, discrimination, social exclusion, and personal and collective devaluation, and it is in favor of social and cultural recognition; in other words, it refers to the emblem of *'social justice, cultural recognition and new human rights'* and, therefore, it refers to an encounter for a sociocultural democratization. The gains in this dimension are connected with the next one.
- The politically necessary conditions to create social subjects or actors for emancipation processes: which include struggles against lack of citizenship or lack of empowerment or participation in the political processes; in other words, this dimension refers to the emblem of the

need for 'social subject autonomy, democratic public sphere and institutional participation', in other words, to act in the name of a political democratization.

The first dimension, the material demands of daily life, is the major mobilizing factor of the local movement's bases for the formation of the so-called popular movement. For instance, the National Forum for Land Reform and Justice in the Countryside (FNRA), which includes fifty-four national organizations, besides its historical struggle for a massive land reform, demands the legalization of land for indigenous and marooned (or *quilombola*) populations, resources for production, retirement rights, public health, rural education, and so on, which require direct negotiation with the state and governments.

More recently this forum joined a broadly political campaign, networking with several rural and urban movements such as MST and others, asking for a law that imposes a limit on the size of rural properties. Behind this political act, the movement is constructing a new conception of 'nation', where the three dimensions above would be present, through a policy that combines structural changes (such as the limitation on property size) and public policies for social inclusion of the historically excluded or discriminated people of our nation: the Indians, the descendants of slaves, and the poor of rural and urban areas. This can be illustrated by the words of a member of the Forum for Defense of Indigenous Rights:

> The Indians are considered to be an embarrassment for development and they [the country's elites] have a saying: "too much land for too few Indians." In fact we have too much land for too few white people. The indigenous land is public land, but it is the cultural ground for these people. Nevertheless, businesses with their capital think that this land should be open to their idea of development. (FDDI interview)

In this regard, several of the thematic forums and civil society networks in the research raised the necessity of a movement's policies against structural historical legacies in our cultural model, such as the culture of slavery, patriarchal colonial culture, inequality rooted in cultural and regional differences, racism, etc. In other words, the three dimensions—material, symbolic, and political—appear to be discursively related. Another question raised by these forums was the necessity of working for the visibility and empowerment of the groups that are socially excluded historically, as is evident from the following words of members of the forum about garbage and citizenship and the forum for black women:

> We need to work for an organization and a policy where the garbage collectors speak for themselves, where they develop a political consciousness and a capacity to insist on their rights. (FLC, interview)

We need to seek visibility of the history of the black people in Brazil and to have recognition as political actors at councils for black people, at public forums, at SEPPIR (Ministry for the Promotion of Race Equality), etc. (FNMN interview)

The search for public visibility and empowerment represents two complementary faces for the politicization of these movements and for their political learning towards institutional participation, as we will see next.

Empowerment and Institutional Participation

"We want the state to promote public policies, but we want to say what needs to be promoted." These are the words of a militant of the National Forum for Black Women (FNMN).

The organized civil society and social movement networks demand a new form of participation in the public sphere that focuses on empowering the historically excluded subjects in the development of full citizenship, and that reflects the principal types of human rights: civil, political, economic, social, cultural, environmental, and heritage, including those of minorities.

In this regard some results of our research are the following. In general terms we observed in these forums that the ideals for equality were related to the principle of social recognition of differences, taking both as the means for policies of equity and social justice. As a result we found that forums and network movements are defending and working for the implementation of a human rights platform[8] that takes into account the following three dimensions: laws and policies for equality, for the recognition of social and cultural differences, and for the empowerment and political participation of the excluded groups in the public sphere. Within the frame of civil rights, the main struggle is for the observation of human rights already included in our Constitution but not yet applied to many segments of the country's population:

The provision of citizen documents (birth certificate, identity card, voter registration, etc.) is fundamental, because this means promoting citizenship, to protect rights that should belong to every human being and not only to the 'others'. (FNRA interview)

The first claim to citizenship is the right to a birth certificate, but many children do not have the right to go to school, to receive vaccinations, etc., because they do not have this document, and they are excluded from the beginning of their life. (FNPETI interview)

Our principal struggle has been for a change in civil regulations, for example against the principle of tutelage[9] for the Indians by the state [it was mentioned as still being applied in practices by FUNAI/National

Foundation of Indians], because they are considered to be relatively incapable. (FDDI interview)

This situation of exclusion in the plan of civil rights is the result of state policies that have historically failed to take into account structural inequality and cultural differences in our country. Within the frame of socioeconomic rights, we observed in the agenda of these networks the defense of rights that should already be considered as universal such as the access to food, health care, land for work, etc., but we were also reminded of the need for the creation of laws that promote structural changes toward broader social emancipation:

> We need to consider the concentration of income as an indignity to human rights. We need to consider the access to land, to production, to food, and the 'minimum income' as human rights. (FDDH interview)

> Acting for public policies in the field of economic rights (such as in the budget monitoring) we are creating a step towards the realization of other human rights. (FBO interview)

The following rights were also mentioned in the interviews: the need for a broad application of the legislation against child labor (FLC), for the legalization of the indigenous and quilombola land, and for the observation of the social function of rural properties (FNRA). And in the line of compensatory rights, the interviewees defended the need for affirmative action policies in the promotion of social equality and cultural inclusion, such as ethnic quotas in universities, women's quotas for political participation, etc. (FENDH). What these social movement networks are proposing is a nation based on social rights for everybody, in their words "a nation that is socially just and politically democratic." This takes us to the next human right.

Considering the frame of political rights, the conflict originating from the dictatorial times between the autonomy of the subjects and institutional participation in the state takes on new configurations nowadays within a more democratic scenario. The organized civil society tries to combine the political autonomy of the subjects with a growing and broader participation in policy making, as a means to develop a more participatory and simultaneously a more radical democracy. The following testimonies illustrate this tendency:

> The state needs to recognize that the indigenous leaders are political leaders of their people and territory, and so they need to be recognized as legitimate political leaders of the Brazilian nation. (FDDI interview)

> In the MST we have a participative methodology that starts at home, in their organizations, from the movement's basis to its coordination.

Changing the participatory mentality at home, in the movement, you create the conditions to participate differently in the public institutions and in the struggle for political empowerment. (FNRA interview)

These forums advocate the need to construct the 'legitimacy of participation' of representatives of the historically excluded people of the country. They also realize that the learning process for political participation needs to include the movement's basis. Therefore if empowerment needs to be constructed from educational and political processes developed at the base of the movements, i.e., on the localized organizational level, it needs to be simultaneously deepened through the subject's participation in the political articulation and mobilization at the level of the public sphere. And to have institutional results it needs to be alert to the political opportunities[10] created by the state, as has been done in several Brazilian National Conferences[11] that have increased substantially in number since Lula's government took over in 2003 (see the Graph 5.1 illustrating the expressive growth of National Conferences in Brazil).

Within the frame of *cultural and minority rights*, we found many demands for the incorporation of new human rights. As culture is usually taken just as a legitimate custom or tradition, the forums and the movement's network defend a critical revision of the colonial ideology and practices of slavery. They work with the movement's basis for reinterpretations in the field of cultural reproduction, introducing a rereading of the origin and cultural diversity of the Brazilian people (especially in relation to Indians, and black and mestizo peoples), and a deconstruction of the ideology of 'racial democracy' of the dominant elites.

In this regard they advocate laws for the demarcation of indigenous lands, taking into account the intrinsic relation between territory and culture; in their words, "if there is no territory, there is no way to keep our culture alive"(FDDI). This is also very relevant for indigenous people of the countryside and forests.

The pedagogic role of the forums was remembered in relation to the education for diversity, related to race, ethnicity, social or regional origin,

Graph 5.1   Conferências Nacionais Realizadas (1941–2010). Source: Secretaria Geral da Presidência da República.

gender, generation, special needs, etc. (FNPETI, AMB, FNRA, and others). The political interface constructed between cultural traditions and new values helps the movements to develop associations among the human rights of several generations, such as the need to consider the interfaces between civil, economic, and cultural rights.

Finally, in the frame of *environmental and heritage rights*, the already classical demands on regulation for preservation of the forest and biodiversity appear together with the protest about the social and ecological consequences of agribusiness, transgenic seeds, biopiracy, and so on. In the field of heritage, new human rights are indicated, such as to consider water as a common heritage of all humanity (FNRA), the need for responsible consumption, and the fact that rights nowadays need to take into account the rights of future generations (FBES, ABONG, FDDI). In this sense, ecology is becoming ethically related with quality of life, with new cultural choices, and with rights of citizenships.

THE PARADIGMATIC CASE OF THE MST (LANDLESS SOCIAL MOVEMENT)

We consider the MST as a paradigmatic case because this movement has fulfilled the three constitutional levels of a networked social movement frame (organizational, articulatory, and mobilizing), and the dimensions for its political construction (principles of identity, conflict definition, and project of change) have been developed at all levels of the network. In addition, the most paradigmatic and substantive demands of contemporary social movements are also present in MST. On the one hand they illustrate the singularity of Brazilian social movement networks, and on the other they show some connectivity with the more universalistic ideals of global movements. We will now examine how the MST has been negotiating these demands with the institutional spheres of the state.[12]

Firstly, *the material/emergency demands of daily life* are the major mobilizing factor for the movement's bases, through which they construct the principle of collective identification—the landless—and a political network strategy for empowerment through the formation of the self-called 'popular or mass movement'. Although the search for concrete answers (conquest or legalization of land, resources for production, retirement rights, public health, rural education, etc.) requires direct negotiation with the state and governments, civil disobedience by squatting on unproductive farms, together with manifestations in the public sphere, constitute their main political strategy. Only after constructing public visibility does the movement work for negotiations with the state, when it has faced the challenge of creating pressure without being co-opted. Under the Lula government this has become more difficult due to the strong ideological identity that the movement had with the former candidate. During President Lula's first administration, certain guidelines for struggle were suspended in order to avoid harming the electoral process of reelection, although afterwards this was partially reviewed. Thus, at the beginning of the second mandate, the social movements' bases began to perceive that the acquisition of land was moving even further out of reach. Encampment became the most possible realization. The political justifications of their leaderships were called into question. Consequently, the MST, through and with the National Forum for Agrarian Reform and Justice in the Countryside (FNRA), relaunched a national campaign for the legal limitation of land properties (Vigna 2007).

Secondly, *the praxis of reevaluation of symbolic and political meaning inside the movement and within society* became a fundamental moment of political formation and for the attainment of cultural recognition of the movement's subjects. The political and symbolic nexus established at the three levels of the MST's network were fundamental in this process:

- At the *organizational level,* the movement focused on educational practices for symbolic deconstruction and reconstruction related to the politics of identity—*the landless as a political actor*—in agreement with the movement's logic—*the struggle for land in connection with the construction of a new project of society*; also for the formation of autonomous subjects, but with a collective role in the process of social change.
- It is on the *articulatory political level* that the collective political identification is consolidated and where the national politics for the movement's networks are discussed and consolidated; it is also where the subjects learn to respect the differences between political and ideological options, together with the regional, ethnic, age, gender, and other cultural differences, as has been occurring in the national forums.
- It is at *the level of mobilization in the public sphere* that we see the movement's search for political visibility and for broader public recognition, seeking political empowerment as a condition to open channels of negotiation in the government sphere. This constitutes what is being called a "protest or warring movement" in the sociological literature.

Thirdly, networking with other popular movements, the MST moves politically and ideologically towards a new project of society and towards changing the hegemonic project of our society. Regarding the empowerment of a broader 'popular or mass movement', it searches for connectivity with other Brazilian movements and civil organizations such as the Via Campesina, the Affected by Dams (MAB), the indigenous, the *quilombolas*, the homeless (MTST),[13] and other movements; and through many civil forums and political articulations such as the Center of Popular Movements, the National People's Assembly, the Popular Referendum, the Brazilian Social Week, the Cry of the Excluded, the Association of Brazilian NGOs, the Brazilian Social Forum, etc. On the world stage, more organic networking is done with the transnational Via Campesina, the Continental Cry of Excluded, ALBA (Bolivarian Alternative for the Americas), and Internationals Caritas and FIAN International (Food First Information and Action Network). This is also achieved through its actions on the World Social Forum.

The major current debate concerning the construction of a 'new national project' contemplates the following aspects and political ideals: In relation to the modernizing model in agriculture, this project opposes the marketization of land reform, the type of agribusiness, the transnational companies that control seeds, and the widespread agrarian production and commerce, as well as slave labor and other kinds of subordination in the country. MST also proposes agriculture focused on the internal market, with respect for the environment and stimulating agrarian cooperation and the autonomy of labor, as in family farming.

In relation to land ownership, MST proposes a limitation on the size of property, as mentioned in the document "Earth Charter," which, for MST and according to the FNRA, is what brings "total unity to all rural popular movements." Here there is support for the expropriation of all very large properties and those that belong to foreigners, banks, and those who practice slavery. There is also the struggle for the state's recognition and legal delimitation of all indigenous territory and remaining *quilombola* communities.

Finally, the construction of a new national project faces some challenges in the internal organization of the movements' network, such as building understanding around common struggles and considering at the same time the heterogeneity of collective actors, historical roots, and referential political fields (religious, political-partisan, classical and renewed Socialist left-wings, and other outspreads), expanding their political networks with political actors from other referential territories (urban, Latin American, and globally). The tension here results from fractions that seek unity with certain political costs, with the objective of building a unified front or political counterforce against the system, in relationship with other fractions that want to construct movement networks with more open guiding principles for common action and with more space to respect the differences and diversities of specific struggles. The first stream advances towards a movement conceived as a political organization, whereas the second tends to keep itself as a movement in an open process, in other words, as a networking social movement.

CONCLUDING REMARKS

Taking into account the research results presented up to now, I would like to mention three conclusions regarding the emancipation processes of Brazilian social movement networks nowadays. Firstly, the growing number of civil organizations and thematic social movements representative of historically excluded or discriminated groups of Brazilian society (such as the emblematic movements of garbage collectors, the solidarity economy, black and indigenous people, women and especially black women, struggles against slavery and child labor, landless and homeless people, racial and gender discrimination, etc.) has renewed popular political agendas, demanding simultaneous changes in public policies for social equality and the recognition of cultural differences. Although the demands for greater material equality are found acceptable by most social movements, the admission that cultural differences produce discrimination, which also supervenes upon social inequality, is a stronger agenda among the new cultural movements, such as in feminism, ethnic movements, and so on.

Secondly, the existence of broader network organizations like the ones organized in social movements, together with all sorts of social forums, may promote dialogues between the thematic singularities of each civil organization, representing the plurality of social voices. These dialogues have produced new perspectives on emancipation in that they help to construct the necessary political understanding about the connection between the three following levels of the movements' formation: first, the material demands, resulting from precarious material conditions of life; second, the symbolic conditions in social reproduction, expressing the connecting forms of exclusion and discrimination; and third, the political pedagogic processes towards the construction of the movements' autonomy combined with their participation in a democratic public sphere in policy making, and representing their social roots.

Thirdly, another important political fact is that these organizations are networking the movements' basis with regional and national (and sometimes global) movement organizations through mediators, forums, and so on, not only working for the movements' empowerment and visibility through public protests and manifestations, but also aiming at the establishment of new rules and policies for the universalization of classical human rights (such as in the field of civil rights, work regulations, quality of life, education, and health) and for the implementation of new human rights, associated with cultural differences (such as those for compensatory policies and affirmative actions). Finally, they attempt to defend and promote popular institutional participation as a path to achieving these aims and to empower the social groups that historically have suffered deprivation, social exclusion, or cultural discrimination.

NOTES

1. These movements were inspired by virtual mobilizations, and afterward they inspired other mobilizations such as the Indignation Movement in Spain and other European countries, Occupy Wall Street in the US, similar protests in the UK, and several manifestations in Latin America.
2. See also the roundtable held by UNESCO on June 21, 2011, at the organization's headquarters on "Democracy and Renewal in the Arab World," which

provided a forum for the diverse voices of Arab society and initiated a truly interactive and dynamic debate, involving both the youth voice and analyses from international experts, as well other countries' experiences of transition to democracy, with emphasis on the range and diversity of viewpoints" (UNESCO 2011: 6).

3. About the concept of discursive articulations see Laclau (2006, 2011) and Mouffe (2003).
4. Title of a project that I have just concluded, "The Multiple Faces of Social Exclusion" (supported by CNPq, UFSC, UNB, and UFMG); see Scherer-Warren (2012). Each one of these forums represents hundreds of local or national civil organizations.
5. The World Social Forum was a very relevant reference for the majority of the forums or networks of the research.
6. Organizational frame refers to the types of organizations, the forms of their interactions, and the respective political meaning of their collective actions, that promote a movement's logic for their praxis. See also Kelly Prudêncio (2011).
7. On this conception of social movements, see Touraine (1997), Melucci (1996), Castells (1997, and some of my previously published works (2005; 2012).
8. In Brazil, a network of several civil organizations is applying the DhESCA Platform (The Brazilian Platform for Economic, Social, Cultural and Environmental Human Rights), with the aim of coordinating the production of the alternative reports on social exclusion and citizenship, for the execution of the International Pact of Economic, Social and Cultural Rights (PIDESC), and as a chapter of the Inter-American Platform for Human Rights, Democracy and Development (PIDHDD). This platform is a national articulation of movements and civil society organizations that develops actions to promote, defend, and repair the economic, social, cultural, and environmental human rights, seeking the invigoration of citizenship and the radicalization of democracy. See Scherer-Warren (2012).
9. The principle of tutelage is no longer accepted by the Brazilian Constitution.
10. See Tarrow (2011).
11. Human Rights, Promotion of Racial Equality, Politics for Women, and other conferences related to demands of the social movement's networks.
12. See other and further developments in Scherer-Warren (2007).
13. MAB/Movimento dos Atingidos por Barragens; MTST/Movimento dos Trabalhadores Sem Teto.

REFERENCES

Castells, Manuel. 1997. *The Power of Identity, The Information Age: Economy, Society and Culture.* Vol. 2. Cambridge, MA: Blackwell.
Laclau, Ernesto. 2006. "Inclusão, exclusão e a construção de identidades." In *Inclusão social, identidade e diferença: perspectivas pós-estruturalistas de análise social,* edited by Aécio Amaral Jr. and Joanildo Burity, 21–37. São Paulo: Annablume.
———. 2011. *Emancipação e diferença.* Coordinated and technically revised by Alice C. Lopes and Elizabeth Macedo. Rio de Janeiro: EdUERJ.
Melucci, Alberto. 1996. *Challenging Codes: Collective Action in the Information Age.* Cambridge: Cambridge University Press.
Mouffe, Chantal. 2003. "Democracia, cidadania e a questão do pluralismo." In *Política & Sociedade: Revista de Sociologia Política,* 11–26. Florianópolis: UFSC.

Munanga, Kabengele. 1999. *Rediscutindo a mestiçagem no Brasil: identidade nacional versus identidade negra*. Rio de Janeiro: Vozes.
Prudêncio, Kelly Cristina de Souza. 2011. "Prudêncio." In *Mídia e esfera pública contemporânea: ação política na internet*. Congresso da SBS, Curitiba. http://www.sbsociologia.com.br/portal/index.php?option=com_docman&task=cat_view&gid=194&Itemid=171 (accessed October 2011).
Santos, Fabiano G. M., and Thamy Pogrebinsch. 2010. "Contra a falácia da crise institucional." *Insight Inteligência* 49: 100–105.
Scherer-Warren, Ilse. 2005. "Redes sociales y de movimientos en la sociedad de la información." *Nueva Sociedad* 196 (March/April): 77–92.
———. 2007. "The Social Movements' Politics for the Rural World." *Estudos Sociedade e Agricultura* 15 (1): 5–24.
———. 2012. *Redes emancipatórias: nas lutas contra a exclusão e por direitos humanos*. Curitiba: Editora Appris, 2012.
Tarrow, Sidney. 2011. "Global, Conventional and Warring Movements and the Suppression of Contention: Themes in Contentious Politics Research." *Política & Sociedade: Revista de Sociologia Política* 10 (18): 25–49.
Thoreau, Henry David. 1989. *Desobediência Civil: Resistência ao Governo Civil*. Translated by Antônio de Pádua Danesi. Rio de Janeiro: Martins Fontes.
Touraine, Alain. 1997. *¿Podremos vivir juntos? La discusión pendiente: el destino del hombre en la aldea global*. Translated by Horácio Pons. Buenos Aires: Fondo de Cultura Económica.
UNESCO. 2011. *Road Map—Democracy and Renewal in the Arab World: UNESCO supports the transitions to democracy*. Document drawn up in the preparation for the high-level roundtable held by UNESCO on June 21, 2011.
Vigna, Edélcio. 2007. "Governo responde às propostas do FNRA." July 30. http://www.inesc.or.br Jul. 2007.

6 MST's Agenda of Emancipation
Interfaces of National Politics and Global Contestation

Breno Bringel

The Landless Workers' Movement, *Movimento dos Trabalhadores Rurais Sem Terra* (MST), is one of the most studied social movements in Brazil and the world. The movement's research agenda and research on the movement (which do not always converge) include subjects such as education, economics and production, law, health, sociology, political science, history, geography, and anthropology. And themes such as the movement's pedagogy, education in the countryside, political education, subjects of the pedagogical process, its economic, political, and communication strategies, cooperativism and development, the agroecological transition and sustainable agriculture, the criminalization of the movement, human rights, interactions with the state, gender relations and women's participation, youth, the creation and territorialization of the movement, social and collective memory, the mystique, landless identity, and an extensive et cetera that defies systematic categorization.

Many factors lead researchers to analyze the MST, but some of the main incentives are due to the movement's capacity to combine in overlapping layers many dimensions of the social, cultural, economic, and political issues as well as its territorialization and size (it is considered the largest social movement in Latin America) and longevity (it was founded in 1984). Other aspects that have been increasingly emphasized are the current reconfiguration of the MST (which goes far beyond the struggle for land and agrarian reform) and its 'global dimension' and influence in events, arenas, networks, and social movements all over the world. However, unlike previously mentioned dimensions, the internationalization of the MST and its global dynamics have not yet been analyzed in a systematic manner, unlike other significant social movements that had a great global impact, such as Zapatismo (Olesen 2004; Rovira 2009) or transnational networks such as Via Campesina (Borras 2008; Desmarais 2007; Vieira 2011). Neither has there been a systematic monitoring of the transformation of the movement's practices and agendas in the past few years. In this chapter, my intention is to partially fill this analytical void by discussing the reconfiguration of the MST agenda vis-à-vis the continuous tensions among the scales of its interests (from the local to the global, including all intermediary arenas, especially the national), the actors/subjects involved, and constructing new emancipatory horizons.[1] This chapter thus strives to offer an analysis of the reconfiguration of the MST's agenda while taking into account the movement's history (reconstructing its history through different cycles of contestation) and space (searching for interfaces between its different scales of action, emphasizing the tensions between national political and global disputes). My central argument is that an especially productive (though not exclusive) method to understand the global projection of the MST and the reconfiguration of its agenda is to analyze its different cycles and consider both the dynamics and processes of internationalization of the movement as well the internalization of supranational exchanges within its political struggle in Brazil.

CRISES, CONTESTATION CYCLES AND EMANCIPATIONS: THEORETICAL-METHODOLOGICAL PATHS

During the last years of the Lula administration and the first years of the Dilma government, many interpreters and media suggested that the MST was in the midst of a 'crisis' and near its 'end'.[2] Several reasons were given: the growing Brazilian economy, the demobilizing character of social policies such as the Bolsa Família, which had pulled families out of extreme poverty but which also removed them from social struggles, severely undercutting the recruitment of new members; the contradictory relationship with the Workers' Party and a supposed proximity with governmental policy; and fragmentation and internal division, among other issues. Going beyond media manipulation and the usual portrayal of MST actions as criminal, a detailed analysis of the reconfiguration of its agenda and actions, inside and outside the national arena, will allow for new interpretations of its emancipatory projects as well as contribute to deconstructing problematic opinions concerning the 'defeat' and the 'end' of social movements.

It is a rather complex endeavor, in theoretical-methodological terms, to speak of 'victories' and 'defeats', 'success' or 'failure' when analyzing the results of a movement's actions. The MST formally came into being in 1984 as a social movement that fought for land and agrarian reform in Brazil, to this day still the country with the second-highest rate of land concentration in the world according to 2006 figures provided by the Brazilian Institute of Geography and Statistics (IBGE) and the UN Food and Agriculture Organization (FAO), second only to Paraguay. It became organized and territorialized at the national level, and even though it has not achieved its main objective (quite the contrary, agrarian reform seems to be less present in political discourse today than when the movement began), it would be questionable to speak of the movement's 'defeat'. If that were true, how to explain all its camps and settlements that continue pressing for the democratization of the land? And how to explain its educational capacity, which includes formal and nonformal arenas, itinerant schools, and even universities? This is also true for initiatives in the area of communication, production, health, etc.

Many social movements dissolve when their demands are met. Others reinvent themselves continuously, go through different cycles of mobilization and demobilization, create new demands and agendas, allies and enemies, and end up contributing in the long and medium run to broader processes of social transformation in a society, as is the case with MST. In the literature about social movements, the development of the notion of 'cycles of protest', and in a broader sense 'cycles of contention' (Tarrow 1998) has allowed for the understanding of collective actions in the immediate political context, rendering the idea of the successes and failures of movements problematic. The starting point for these analyses is the fact that movements come and go; they oscillate according to the dynamics of conflict and social struggle, the search for the 'new' or the replacement/conservation of the 'old', which results in a resignification of the forms and repertoires of collective action and the frameworks for the interpretation of reality.

Cyclical analyses of contentious collective action generally seek to provide answers for a period that spans the arch beginning with mobilization, followed by the intensification of the conflict and its diffusion, and ending with its demobilization or other possible results, such as fatigue, polarization, repression, institutionalization, and, why not, revolution. With this method it is possible to focus not quite on the progression *between* cycles of contestation but, rather, the structure and dynamics *of* the cycle itself, with an emphasis on the rhythm of innovation happening in confrontation methods and the appearance of new referential frameworks for collective action. This can be problematic, because if we endow cycles per se with an absolute centrality, there is the risk of neglecting the nuances of the role played by the social actors within the cycles, as well as their relationships and the creation of identities and different meanings. However, despite the notion that the cycle of contention has been linked to the historical reconstruction of collective actions with the variables having a much more static and linear dimension (Brockett 2005), its analytical potential is considerable as long as it is approached in a more spatial and contextual manner.

According to this line of reasoning, the evolution of social movements and their socio-spatial practices, agendas, and demands can be associated with a cyclical dynamic derived from political readjustments in the processes of material and symbolic redistribution, developing repertoires of contestation within national and supranational spheres. The motivations that cause the rupture points between the different cycles may be responsive to internal or external factors, or a combination of both, but another caveat is in order here: whereas the "transnational cycles of contestation" (Bringel and Echart 2010; Tarrow 2012: 143) are increasingly acknowledged, the separation between national politics and global dynamics is habitually done in a way that is overly rigid, making it difficult to visualize their mutual relationship and the social construction of other spheres in the reconfiguration of collective action and social movements.

This is not merely a question of a persistent state-centric view or an incapacity to overcome the methodological nationalism, but it also touches upon a vision still deeply entrenched in the political and theoretical practices that tends to create narrow divisions between the social and political spheres. This vision is buttressed by an intrastate political theory, which tends to place 'internal affairs' (such as rights, justice, community, duty, identity, and legitimacy) in opposition to 'external affairs' (such as security, war, cooperation, violence, etc.), which would be more appropriate for an interstate political theory (Connolly 2002). According to this logic, as David Slater (2000: 508) reminds us, "social movements are linked to the political domain through their impact on public policies or the priorities of political parties, but any connection to global politics is carried out through the mediation of the internal political system."

I will use the category of contestation cycles in an unorthodox manner in order to analyze the reconfigurations of MST in different spatial-temporal coordinates, emphasizing its agenda and actions in relation to other actors in Brazilian politics as well as its internationalization and global dynamics. The objective is not to separate 'national politics' from the movement's 'external actions' but instead to find points where processes reinforce each other, where tensions and interactions between internal and external dimensions arise in different contestation cycles. The internationalization of the movement and the internalization of supranational references thus appear with greater complexity within the international relationships of MST and contemporary social networks and movements.

By way of a historical reconstruction of MST I suggest that, since its creation in 1984, the movement has gone through four main cycles of contestation:[3] (1) the first cycle (1978–84) of the movement's gestation coincides with the struggle of Brazilian social movements during the democratic transition and the beginning of the movement's internationalization based on its initial contacts with other rural Latin American social movements; (2) in the second cycle (1985–92) the movement begins its territorialization in the national sphere and forges alliances, initiatives, and campaigns in the international sphere; (3) the third cycle (1993–2000) coincides with a greater role of the MST in the fabric of Brazilian social mobilization, a period marked by the Workers' Party (PT) victory in presidential elections and by its international consolidation within a broader transformation of global social activism.

In each one of these cycles, it is possible to identify ruptures in the agendas and political actions of MST that push against, with varying intensity, the boundaries of the movement's national scope of action and allow it to incorporate new meanings of emancipation. Because the main objective here is to discuss the current reconfiguration, I shall focus on the last cycle of contestation in order to discuss three elements I consider to be essential: the redefinition of allies and political enemies, the transversalization of collective action, and the creation of demands with the power to agglutinate local and global grievances. It is important, however, to delve briefly into the previous cycles to map and analyze both the historical construction of the movement's internationalization and some important moments of inflection in its trajectory, all of which are essential to understand its current reconfiguration.

THE SOCIAL AND HISTORICAL CONSTRUCTION OF MST AND ITS INTERNATIONALIZATION: A BRIEF HISTORY

Even though 1978 is assumed as a starting point for the study of cycles of contestation in postdictatorship Brazil, it is necessary to mention some elements of the previous scenario. If the 1960s were marked by an intensely contentious tone in the international scenario with mobilizations springing up in different parts of the world (among them, the anti-Vietnam War demonstrations, the Prague Spring, May 1968 in France, the student protests in Mexico, or the revolutionary and socialist struggles in several peripheral countries), in a national context, the military coup of 1964 interrupted an ongoing process of sociopolitical democratization, which involved significant popular mobilization demanding 'basic reforms'. Agrarian reform was included among them, along with education, tax, and other redistributive reforms that encompassed a new "promise of citizenship" in an unequal Brazilian society (Cardoso 2010: 789–92).

The coup also led to the rupture of three peasant organizations that fought for agrarian reform and ended up influencing the construction of the MST's historical memory: the Peasant Leagues (*Ligas Camponesas*), the Landless Farm Workers' Movement (*Movimento dos Agricultores Sem Terra* [MASTER]), and the Union of Land Workers and Agricultural Workers of Brazil (*União de Lavradores e Trabalhadores Agrícolas* [ULTAB]). These three experiences represented distinct forms of peasant organization (ULTAB and MASTER followed a more party-like model to demand rights for rural workers, whereas the Peasant Leagues emphasized the raising of political consciousness among the peasants so they could fight by themselves, despite the powerful influence of traditional and charismatic leaders) in different parts of the country (ULTAB had greater nationwide influence, present in almost all states, except Pernambuco, where the Peasant Leagues were quite strong, and Rio Grande do Sul, where MASTER had a significant presence). Even though the MST considers itself the direct heir of the historical experience of the Peasant Leagues, all of them represent a memory that is not only historic/temporal but also organizational, thematic, and spatial in the Brazilian rural struggles that appear in the middle of the twentieth century.[4] In all of them the interpretative frameworks for the international reality are mediated by the perspectives and actions of the Brazilian Communist Party (PCB) and the Red Wing of the Communist Party of Brazil (PCdoB), and the frontiers between 'nationalism' and 'internationalism' seem to be accompanied by the importance attributed to the definition of a 'national project' in moments of open disputes for developmental models and formulas, whether they be more reform minded or revolutionary, for undermining or breaking away from the position of dependency.

In the absence of a favorable "structure of political opportunities" (Tarrow 1998) during the dictatorship, part of the Brazilian left chose armed struggle, suggesting that guerrilla groups and collective insurgent action could itself create favorable conditions for social transformation and mobilization. Other militants, be they factory workers, union members, or rural leaders, ended up in exile. This was the fate of a good number of Peasant League leaders. Many of them, such as Alípio de Freitas, fled to Mexico and then Cuba, where they completed many intensive courses on guerrilla training before coming back to Brazil. Military-ruled Brazil also received exiles, due to the military coup in Chile, the beginning of the military dictatorship in Argentina, and several instances of repression during Stroessner's dictatorship in Paraguay.

Among these Southern Cone exiles, there were some peasant leaders, many of whom contacted unions, social rural movements, and religious organizations in Brazil, which were the seedlings of the MST. This is an important point, and I believe that exile (that of Brazilian peasant leaders in other countries and vice-versa) was one of the first moments of formation and diffusion of contemporary global rural activism. Within networks and human flows, transnational exchanges allowed for the creation of an informal network of contacts and trust, which became crucial for the planning of later protests, campaigns, and initiatives. We can mention the example of peasant leader Magui Balbuena, an important reference for the peasant struggle in Paraguay, who fled to Foz do Iguaçu during the 1970s, when a massive number of peasants were driven off their lands due the construction of the Binational Itaipu Hydroelectric Dam.

In an interview with Malgui Balbuena of the National Coordination of Rural and Indigenous Women, CONAMURI, I found that her relationship with MST goes back much further than with her own movement, and in fact it extended to even before the MST came into being. This is how Magui began to tell me about her experience: "I exiled myself with my partner in Foz and we contacted unions, some peasants, and *pastorais*[5] . . . it was an important experience because it allowed me to live some of the experiences of the Brazilian peasants, like the Itaipu conflict, and deepen my friendship and affection for the Brazilian militants . . . I met many of them again in 1985 when I participated, as a member of the Paraguayan Peasant Movement, in the creation of the Southern Cone Coordination of Peasant Organizations and especially in the decade's end with the Continental Campaign 500 Years of Indigenous, Black, and Popular Resistance" (interview 2010).

These exchanges, a consequence of exile, are important not only because they created informal networks but also due to another crucial aspect: the transition from a pattern of internationalization mediated by third parties (especially parties and unions) to one where internationalization occurs in a more direct and autonomous way, creating its own, more flexible spaces for the convergence of rural movements.

PATHS TO MST CONSTRUCTION AND CYCLES OF CONTESTATION IN POSTDICTATORSHIP BRAZIL

This brief history of MST provides us with a broader view of its creation within the context of the dynamics of national and global struggles. I will now analyze the different cycles of contestation in postdictatorship Brazil and discuss the elements within each cycle that were crucial for the interface between national politics and global dispute and for the redefinition of the MST's agenda.

Cycle I

In the mid-1970s, many groups extinguished by the military coup and the period of repression that ensued after the AI-5 decree started to rearticulate themselves. Visible expressions of protest such as the demonstrations in the wake of the murder of journalist Vladimir Herzog and the emergence of new demands in urban areas sprang up alongside more invisible dynamics of social and political rearticulation influenced by the growing irrelevance of the military and the returning exiles, who had a prominent role in the creation of new interpretive and action frameworks. Democratization, which had not figured as one of the central issues in the agenda of the struggle for resistance, gradually became a common rallying axis, despite enormous differences as to the projects envisioned by different social actors. This scenario of rearticulation of social struggles created a social fabric that led in 1978 to a cycle of mobilizations, with new social and political actors entering the stage (Sader 1988).

Kowarick (1987) recalls how in 1978 Figueiredo was elected president by a fraction of the military apparatus in a context of institutional crisis and political illegitimacy, multiplying antagonisms even within the hegemonic circles. This was also the year, after ten years of working-class repression, of the suppression of the first great strike, initiated in São Paulo's metalworking industry, and its diffusion led to a more generalized wave of resistance against the state apparatus. This cycle, initiated in 1979 with the beginning of the transition process towards formal democracy, culminates in 1984 with the end of military rule and the establishment of the first civilian government. This is an extremely relevant moment for the gestation of the MST, because it was in May 1978 that some landless families, who had been moved to an indigenous reserve in the Nonoai municipality in Rio Grande do Sul, were expelled from the land. Many migrated to Mato Grosso, others went to neighboring municipalities, and about thirty families decided to occupy a part of a *latifundio*,[6] the Sarandi farm (Morissawa 2001: 123). By then these families were not being organized by MASTER— they followed another model for socialization and political reflection, much more in tune with the discussion and articulation promoted by the Comissão Pastoral da Terra (CPT), a central actor during this period, along with the Basic Ecclesial Communities (CEBs) and other sectors associated with the progressive Catholic church. The result is the occupation, in September 1979, of the Macali Farm in Rio Grande do Sul, a crucial moment/place for the gestation of the MST, whose formal foundation happens in 1984.

The scenario created in 1978 was crucial for the development of contemporary Brazilian social movements, and marked the beginning of the articulation among diverse social forces (urban social and labor movements in particular, but also the black, feminist, student, environmental, indigenous, and the embryo of some peasant movements in rural zones) that came together at the end of the cycle with the campaign for direct elections called *Diretas Já*. The more visible and symbolic struggles, like the ABC strikes (coined after the initial letter of the industrial cities surrounding São Paulo that were the site of mobilizations: Santo André, São Bernardo, and São Caetano), happened alongside multiple local experiences in various locations throughout Brazil, with claims for housing, transportation, health, education, and schools, among other thematic axes that made up the 'popular' urban demands, typical of the social movements of this period. References to the ideas of 'novelty', 'historical moment', and 'autonomy' were constant.

Of all the grievances, perhaps it's in the realm of human rights that a greater interaction between national politics and the supranational processes of contestation is most evident. International solidarity does not operate in a vacuum but responds to cultural and political affinity and geographical proximity (Bringel and Cairo 2010); hence the revolutionary processes unfolding in Angola and Mozambique in the mid-1970s had more impact in Brazil—awakening the interest of sectors of organized civil society who had not discarded a potential revolutionary future for the country—than Nyereri's African socialism in Tanzania. However, the most paradigmatic example of the interaction between national politics and supranational contestation is not the solidarity with African revolutionary struggles but the amnesty movement, in which sectors of Marxist bent, of Socialist tendencies, Catholics, and other Christian denominations converged in mobilizations, welcoming exiles, supporting popular Brazilian movements and also Latin American ones. The Brazilian Committee of Solidarity towards Peoples of Latin America was created and fought for asylum for refugees from other countries and for Brazilian prisoners, torture victims, and 'disappeared persons' in other countries of the region.

The exchange of references and experiences between militants and organizations, especially from Southern Cone countries, was crucial for the definition of national political strategies, even though the outcomes were different in each country.

In this context of nationwide social struggles and in different places of the Global South, peasant mobilizations and land occupations in several states started to network and dialogue, with the fundamental support of the CPT. Rural struggles ceased to be local and/or regional and completely fragmented, they became more cohesive, and their shared connection led to the creation of the MST in 1984 and the organization of a movement national in its scope. Support from the progressive wing of the Church also transcended Brazilian frontiers in this moment of gestation of the movement because many religious European, Catholic, and Lutheran organizations,[7] whose members supported the first land occupations in the south of Brazil, also mediated the initial search for economic resources from Europe.

Cycle II

If the first cycle of contestation was marked by the gestation of the MST, the reemergence of social actors in the Brazilian public sphere after years of disarticulation by military repression, and internationalism marked mainly by exile and defense of human rights, the second cycle of contestation represents a deepening of the first with respect to the structuring of collective actions and demands, the definition of articulation practices, the delimitation of some actors formal organization, the upward scaling of its projects, and the capacity to affect the direction of national politics. The dynamics and intensity of mobilization were still high, however, unlike the previous cycle, characterized greater visibility of the social struggle; the scenario that emerged in the Sarney government and lasted until Collor de Mello's impeachment gradually moved to a mixed pattern of negotiation and conflict.

This cycle was also marked by struggles for social rights and the demands for popular and Socialist democracy and sought to unmask the hegemonic platforms of constitutional and institutional reforms of the regime. The social urban movements are renewed, and "new social movements" (Scherer-Warren and Krischke 1987) aligned themselves with specific, self-identifying ethnic and cultural demands. Rural movements reemerged on the public stage with increasing influence, and MST assumed a prominent role in articulating the social struggles in the countryside.

In fact, the beginning of this second cycle coincides with the 1st National MST Congress, which assembled in the state of Paraná with, according to MST's records, 1,600 delegates from all over the country. Under the heading "Without Agrarian Reform There Is No Democracy," the movement sought to reintroduce agrarian reform in political discourse in a context of mobilizations aimed at the Constitutional Assembly.[8] In this first Congress delegates discussed strengthening MST's national organization by bring-ing together peasants and isolated movements in the country. They chose not to ally themselves with the new government and continued to use the occupation of unproductive large farms as the main tool of their struggle. Educational and agricultural cooperation also make an appearance as pri-orities in the training and construction of social links among militants. During this period, the Workers' Party (PT) and the *Central Única dos Trabalhadores* (CUT) become strategic allies of the MST in the rural and urban fronts to help expand the scope of the popular struggle.

The hope attached to the 1st National Congress on the National Plan for Agrarian Reform (ONRA) was quickly dashed by the significant influence of the Ruralist Democratic Union (UDR), the movement's historical enemy, in the new political scenario. Facing the failure of agrarian reform in the assem-bly, the movement decided to change its motto in 1989 to "Occupy, Resist, and Produce," and managed to increase its strength significantly in terms of numbers of occupation, mobilized families, and national presence.

With regard to international articulation, the big priority of the MST in this cycle was coordination with similar movements in Latin America. This supposes a two-pronged process: firstly, a self-definition of the movement itself and its characteristics; and secondly, establishing identification with similar movements in neighboring countries. This double effort is evident in the 1st National Congress, when the MST sought to define its national identity while inviting several peasant leaders from the region. According to Dulcinéia Pavan, a member of the MST's International Affairs Secretariat, speaking at the European Congress of the Friends of the MST in 2001: "We wanted to know how they were trained in popular mobilization, themes of production, etc. . . . at first we had some difficulties, because most of the organizations were linked to political parties, which made us suspicious. But it was also a thrilling moment, since after the peak of the revolutionary movements, we were very interested in the struggles of peasant movements, especially in Central America: El Salvador, Nicaragua, and Guatemala, but also Peru, Colombia, Ecuador, and Cuba."

Protests against foreign debt payments and structural adjustment poli-cies contributed to creating collaborative campaigns among the region's rural movements. The most important of these continental initiatives was the Continental Campaign "500 Years of Indigenous, Black, and Popular Resistance," launched in 1989 to challenge the official and celebratory nar-rative of the fifth centennial of the Spanish conquest and the 'discovery' of America, which would happen in 1992. During its three years of exis-tence, it was central to establishing a more stable and lasting coordination between the rural Latin American movements. The active participation of MST managed to overcome the historical barriers that separated Brazilian social movements and their Latin American counterparts.

The 500 Year Continental Campaign coincided with the fall of the Ber-lin Wall, which was more than a symbolic coincidence, since under the motto "Unity in Diversity" and with a perspective that diverged from certain practices of so-called real socialism, historical convergences started to appear among the social movements of the region without third-party mediation, be they parties, unions, or other actors. Thus, rural movements managed to create their autonomous space for articulation, in which the previous model of mediation existed alongside a new agenda of direct internationalization. This was a point of inflection, for there is a new scenario of convergences and solidarities in the regional space where the emphasis is no longer on finding new members and creating structures for their coordination, and in which attention turns to the mechanisms and dynamics that allowed for a fluid interaction between different movements and organizations, with a much more flexible articulation.

At the national level, the 1989 elections and Collor's victory opened up a new scenario where the MST continued its organizational and territorial expansion.[9] The corruption scandals that plagued the Collor administration and eventually led to his impeachment became the common thread in protests and citizens' demands. Whereas the moral and ethical discussion became more important to a broader network of actors and ideologies, the end of this second cycle was a sign of things to come. The UN Earth Summit, Rio-92, is emblematic of this change because in this event the 'new social movements' (especially the environmental ones) converged with popular and community organizations and many national and international networks and NGOs.

Based on the debates, dynamics, and interpretative frameworks of participating actors, the event ended up displaying two tendencies of social mobilization dynamics that would exist alongside each other in the next two cycles: a combination of the desire for rupture of some social movements, who still believed in conflict, and the more cooperative action of actors with a higher degree of institutionalization such as NGOs, who tend to act within the political system and its rules and institutions. Therefore, the transition between the second and third cycle opens up a new scenario of mobilization in which institutionalization and internationalization of collective actions are strong tendencies, even if they do not always operate in tandem.

Cycle III

In the third cycle, we find a greater pluralization of social actors and their relationships to the state as well as new sociopolitical and economic scenarios in the national and global spheres. It is a cycle abounding with paradoxes. Right when many social movements, especially rural ones, were becoming stronger in Brazil, much national literature discussed their crisis, dissolution, and even their "end" (Hellman 1995). So today is not the first time the 'death' of social movements in Brazil has been announced. What in fact happened was a relocation of the analytical lenses of many researchers towards other subjects (towards civil society, in a broad and often amorphous context, encompassing participatory democracy, the public sphere, etc.) and actors (especially NGOs and the so-called third sector).

Indeed, during this cycle there was a change in the pattern of social mobilization, which tended towards a greater institutionalization and an emphasis on negotiating rather than conflict with the state, which also led to a powerful process of cooptation of social leaderships by local governments and institutions that lasted until the end of the cycle. However, social movements did not disappear as if by magic and neither did they become completely institutionalized. It is important to understand the intangible and more invisible dimensions of social movements in a time of decreasing public awareness as well as a more tenuous boundary between state and civil society in analyzing social actor's trajectories of activism. At this time, social movements acquired a new form of expression; they sought new repertoires, organizing methods, and strategies (which includes even occupy certain government positions); more structured internal and communicative methods—invisible to an external observer—and demands in a situation radically different from the previous cycle. The tendency to seek alliances and articulations transformed the networks into a new organizational and communication morphology, which contributes to deepen the construction of action plans that go beyond the national, reaching regional and global spheres.

In other words, the very composition and configuration of social struggles is altered: popular social movements, typical of the previous cycles, enter a period of retreat and reflux, working more on internal and daily issues of neighborhoods and communities than on national politics. They must also confront the turbulent reconfiguration of cities, the restructuring of the labor market, and the increasing complexity of the urban and the social issue. In general, the intensity of the urban conflict in most urban mobilizations diminishes, and at the same time, in the local sphere, there is an emerging citizenship participation in public policy. The labor union movement had little or nothing in common with the movements of the previous cycles, becoming less relevant as a tool for worker representation and class struggle while many trumpeted the arrival of a 'new era' of participation. Meanwhile, on the opposite end, rural social movements became stronger, territorialized, and one of the main contestation forces. Thus, the focus of the struggle moved to the rural areas and so does its criminalization and repression, which, nevertheless, resurfaced in a much more selective and refined fashion than during the dictatorship or the first cycle of contention (Bringel 2011).

This brings us to a second paradox: whereas local governments, organizations, and national and international institutions increasingly demand 'civil society participation', more combative social movements and radical forms of protesting are met with repression and criminalization. A certain kind of political participation is welcome (one that fits in a more systemic and reformist framework, which can legitimize government initiatives of poverty reduction, in accordance with directives of the World Bank, IMF, and other agencies), whereas all other types that are perceived as threats are criminalized.

The actions of the government's repressive apparatus (in combination with actions carried out by landowners) led to several murders and massacres in this cycle, including the Corumbiara Massacre on August 9, 1995, and the Eldorado dos Carajás Massacre on April 17, 1996. In 1996, the Via Campesina, which became, in this cycle, the main transnational network for the articulation of global rural movements, established April 17 as the International Day of Peasant Struggle. This date also became a reference for the *Jornada Nacional de Lutas*, launched in 1999 by MST and other social actors, and became known as 'Red April'.

The MST's centrality in the articulatory organizational and oppositional dynamics was significant during this cycle, in both the national and international spheres. In the national sphere, the cycle begins in January 1993, with the approval of the Agrarian Law, which regulated the disappropriation of lands for agrarian reform. One of its victories was the reaffirmation of an issue that had already been set in print in the 1988 Constitution (the social function of land), although another important issue was the inclusion of a guarantee that the landless would be settled in the regions where they lived. In the following month some members of MST went to Brasília to meet Itamar Franco. It was the first time a president opened the Plantalto's Palace doors to the movement. However, this new opening up to dialogue and the adoption of the Agrarian Law did not move forward nor did it translate into a fluid relationship between the movement and the state.

In July 1995 more than 5,000 MST delegates met once again in Brasília to participate in the movement's 3rd Congress, which by then had established a foothold in twenty-two states. The central challenges were still internal strengthening, developing a new agrarian program, and deepening relationships with other social sectors in the country and peasant organizations in Latin America and elsewhere. The Congress takes place during a sensitive economic-political period for the country and the movement as the consequences of neoliberal hegemony were starting to become apparent. It was during this decade that the MST became nationally consolidated and converted into the main social reference point in the opposition against the neoliberal project. The data contained in the "Conflict in the Countryside" report, collected by the CPT (from 1997 to 2000), show how from 1996 to 1999 there was a considerable increase in MST land occupations. In 1996 there were 398 land occupations; 463 in 1997; this figure soared to 599 in 1998; and in 1999 there were a total of 593.

It was during this cycle that the MST managed to become closer to other social actors as well as society itself. Maybe the most emblematic example of this rapprochement was the march that began in February 1997 and brought 1,300 MST militants from different places to Brasília,

where they met for a massive demonstration against the FHC administration. The well-known verses of the Spanish poet Antonio Machado,
Caminante no hay camino; se hace camino al andar, strike a chord in
marches like these, because perhaps more than the destination, what is
important is the path and the journey, the process and the stops along the
way that helped the movement to connect and spread their demands to a
greater audience.

In the supranational arena, we have the creation of the Via Campesina
in 1993, a movement that today comprises 150 organizations and social
movements from seventy countries in Africa, Asia, Europe, and the Americas; and in the regional arena, the creation in 1994 of the Latin American
Coordination of Peasant Organizations (*Coordenadora Latino-americana
de Organizações Camponesas* [CLOC]), which in practice acts as the
dynamic engine of the Via Campesina in Latin America. In Europe the
'MST Support Committees', also known as 'Friends of the MST', were created during a period that can be regarded as a new cycle of international
solidarity with Latin America. Unlike the classic internationalism of the
nineteenth and part of the twentieth century, and the various expressions
of solidarity towards the revolutionary processes and social transformation
in Central America and the Caribbean in the second half of the twentieth
century (Cuban revolution, Nicaragua's Sandinistas, etc.), Friends of the
MST emerged, influenced by the Zapatista movement in Mexico and with
a renewed focus and set of ideals. This solidarity is born out of a direct
connection to the movement and not based on a 'national project', because
these groups do not expect to occupy the state, change the world, and lay
out the path ahead, but instead expect to change 'their world', dissemi-
nating new social practices and alternative ideas. Both the Zapatistas and
the MST became important references for a new generation of militants in
Europe and in other parts of the world, who seek to create societal alternatives where emancipation seems to displace, or at least become associated
with, revolution as a key word.

As a parallel development, the multiplication of networks, movements,
campaigns, and transnational spaces for convergence that sought to join
forces to create alternatives to the hegemonic neoliberal globalization led
the MST to carve out a proactive role in a myriad of initiatives that have
contributed to the contemporary reconfiguration of internationalism in
Latin American and the World. Some examples are the *Grito dos Excluí-
dos*, the World Social Forum (WSF), the Continental Campaign against the
Free Trade Area of the Americas (FTAA), and subsequent creation, already
in the next cycle, of the Social Movement Articulation for the ALBA. The
inclusion of global themes became more in the MST from this moment on,
as it became clear during the 4th National Conference in Brasília, held in
August 2000. Some topics that stood out were the fight against genetically
modified foods and multinational corporations, the actions against international organizations such as the World Trade Organization, the World

Bank, and initiatives like the FTAA, and also the consolidation of April 17 as a reference for the internationalist struggle of the land.

New Agendas and Emancipatory Projects: MST during the PT Government and the Anti-globalization Struggle

After analyzing the social and historical construction of the MST and its internationalization as well as the different cycles of contestation that coincided with its gestation, creation, and consolidation vis-à-vis national and global struggles, we are a in a better position to examine the agendas and perspectives of the movement in the last decade. This last cycle of contention, which begins with a wave of antiglobalization mobilizations and with the 1st WSF in 2001, is characterized by a complex interdependence between the state and society; by the contradictions, comings, and goings between the MST and the government; by the greater interaction between types of political action; by a mixed pattern of mobilization that combines protests and proposals, cooperation and conflict; and by the authorities' reactions, which varied between negotiation, cooptation, and repression. With a scenario considerably different from the one of the previous cycle in both Brazil and neighboring countries, the social movements became the protagonists in Latin America at the turn of the century, forcing a deepening of an always incomplete and impartial democratization, seeking alternatives to neoliberalism, and creating critical synergies with leftist governments in the region, many of them elected thanks to social mobilizations.

Porto Alegre is an excellent metaphor for this cycle. We have here the participatory budget as one of the clearest examples of local participatory policies and of the tensions and convergences between social movements, parties, governments and actors, and institutional practices. Porto Alegre also hosted one of most significant spaces for convergence between networks, organizations, and social movements from all over the world in the first decade of the century, the World Social Forum (WSF). The global and the local thus emerge as related and mutually constitutive spheres of social activism. The motto "Another world is possible" became a symbol of the rupture with the Washington Consensus and the need to create new interpretative models for social transformation.

As a space for the convergence of resistances of multiple places and scales, and of debates and proposals, the emergence of the WSF channeled the synergies between different struggles against neoliberal globalization, which were already being coordinated during the previous cycle. The Seattle protests in 1999 managed to paralyze the WTO and delegitimize its authority in regulating global trade. The valorization of diversity and the plurality of voices; the opposition to totalitarian and reductionist views of the economy, of development, of history, and of social change, and the deliberative character are among the main characteristics of this space for the exchange of experiences (Bringel and Echart 2010; Pleyers 2010).

Porto Alegre's 'two faces' seem to be showing signs of fatigue, due to the weakening of the WSF and other big global conferences as well as the limits of the institutionalized cycle of participation in different parts of Brazil. However, in a scenario of multiple systemic crises (financial, but also ecological, energy, food, etc.), the reconfiguration of global power, and greater regional and thematic decentralization of struggles, we see the consolidation of transnational feminist, environmental, peasant, indigenous, and other networks around thematic axes such as housing, food policy, education, or free trade with the important participation of the Brazilian civil society and the MST.

Even though internationalism and transnational exchanges have been a part of the MST since its gestation, it is only during this cycle that the flows and transnational interactions influence a profound reconfiguration of the ideas and agendas of the movement in all its scales. The internationalization of MST became consolidated in the last few years through the confluence of different plans of action: the articulation with other rural social movements in different regions of the world; the political solidarity of the Support Committees in the center of the world system; the political-economic cooperation with several actors; the participation in networks, international events, and transnational spaces of convergence. All these plans depend on processes of convergences and sociohistorical articulations, on cultural affinities, and on a series of personal and impersonal, symbolic and material, formal and informal elements, which connect the networks to the territories and localized struggles, and vice-versa.

In the case of MST, there is a creative tension between the broader networks (such as the CLOC and Via Campesina) and transnational flows (of people, information, and education) within the reconfiguration of contemporary internationalism and its own agenda for emancipation. Similarly to the previous exile case, the brigades are a distinct element of the human flow that allows MST militants to interact with movements and realities of other countries and vice-versa (for example, the militant brigade in Haiti and the Haitians who interacted with the MST). The communication and education/training spheres are examples of MST's central themes in the national sphere that were translated into the international scenario, on the premise that without education and training (of members, but also technical knowledge about seeds, agroecology, etc.) and without the construction of a powerful communication basis (both internal and external), it would be nearly impossible to advance the diffusion of the ideas of the international peasant struggle.

This significant increase of the MST's global projection coincided with Brazil's global projection enabled by Lula's aggressive stance in cultural and foreign policy that put the country on the international map. I have examined elsewhere (Bringel 2006) the contradictory character of the relationship between the MST and the PT, also within the movement (especially among some leaderships, between the leadership and the members, and between different locations and regions where the movement is present). Initiatives like the Popular Consultation are encouraged by the MST in order to ponder the necessity of a new political instrument to counterbalance the PT's 'ideological drift' and the left's diminished transformative and mobilization capacity. Nevertheless, the initial hope of the MST was quickly destroyed when the PT government made it clear that it would bet on agribusiness and commodity exports as central pillars of its global project. The alliance between the PT and the dominant sectors, the approval of genetically modified seeds, the neglect of agrarian reform, and the expansion of the agricultural frontier, among others, led to enormous challenges that forced the movement to reinvent its agenda.

Without neglecting fundamental structural dimensions, such as the redefinition of the agrarian issue, the financialization of capitalism and food, and the productive restructuring of capital, I would like to point out three crucial elements in MST's twenty-first-century agenda: the redefinition of political allies and enemies, the transversalization of collective action, and the creation of what I call 'aggregating demands'. It is important to analyze how the internationalization of the MST led to an internalization in terms of the discourse, ideals, and local/national practices of global dynamics and problems, present in the Brazilian countryside.

THE REDEFINITION OF POLITICAL ALLIES AND ENEMIES

In the last decade, MST has redefined its alliances and adversaries. Its consolidation as the social movement with the largest membership in the country gave it more autonomy with respect to social and political actors, such as the *pastorais,* which during the first cycle had a great influence over the movement. Many of them remained allies, but the relationship is not as strong as it used to be. It is important, however, to distinguish between strategic alliances and tactical alliances. In the former case, we find mainly other class struggles and rural social movements, with which the MST shares affinity, experiences, repertoire of actions, adversaries and common objectives, projects, permanent campaigns, solidarity, and identity links. These actors changed throughout the different cycles, and by looking at the interface between national politics and global contention, it is interesting to observe how the forging of alliances and transnational networks can rearrange the correlation of forces and relationship between the social actors in the national sphere. MST's participation in the International Via Campesina brought it closer to territorialized movements in Brazil, movements it had interacted with but with whom it did not share strategic visions and plans of action as it does now. In other words, when the Brazilian social movements became part of the International Via Campesina their relationships with each other and with the MST intensified in the national sphere. This happened to the Brazilian

Movement of Dam-Affected People (MAB), the Movement of Peasant Women (MMC), the Small Farmers' Movement (MPA), and other actors that interacted with each other under the umbrella of the Via Campesina Brazil, and that today are the MST's main strategic allies in Brazil. This same dynamic is played out in the global scenario with social rural movements associated to the CLOC and International Via Campesina.

In tactical alliances we find less organic and more calculated support, based on collaboration, coalition building, specific campaigns, and advocacy on public policies. The alliance base is broader but also more ephemeral, changing according to the correlation of social and political forces and the social and political conjuncture. These include a broad spectrum of actors, such as basic ecclesial communities, networks linked to the Constitutional Assembly, civil associations, unions, progressive parties, several NGOs, articulatory networks, support groups, and civil society forums (see Scherer-Warren's chapter in this volume).

Regarding the definition of political enemies, there was also a redefinition due to the constant transformation of the models of struggle and conflict. The PT did not turn from an ally into an enemy, but from a strategic ally in the first two cycles to a tactical ally in the last two, although increasing tensions are noticeable. The MST's historical enemies—the large landowners, the ruralists, and the state[10]—are still their main focus but now share the stage with other enemies, especially corporations and international capital. Monsanto, Bunge, and Cargill are now specific targets of protests due their production of pesticides and GMOs, and as the main proponents of the view that agriculture and food are a business and not a right.

THE TRANSVERSALIZATION OF COLLECTIVE ACTION

European debates on new social movements, clearly not a homogeneous trend or paradigm, suggested there was a division between material and postmaterial struggles, typical of postindustrial society. This division was never accepted wholeheartedly in Latin America, because it would be difficult for us to diminish the importance of popular movements and material struggles (for the land, housing, etc.) to accommodate movements with more symbolic and cultural demands (pacifism, environmentalism, feminism, among others). We saw in the first two cycles of contestation how the MST dealt with the scenario of Brazilian social movements characterized by ethnic, racial, gender, and environmental struggles. I would like to suggest that a fundamental change happened in the third cycle and became consolidated in the last one: symbolic and cultural struggles were gradually incorporated by the Brazilian social movements and combined with eminently material struggles, and the MST is the most paradigmatic example of this process.

In the last decade the idea that gender and environmentalism should be transversal issues and incorporated by different social movements and struggles has become widely accepted. Hence feminist and environmentalist struggles cannot be advanced only by movements with specific grievances. All social movements should adapt their interpretative models and actions to incorporate these issues into their causes. In fact, this constitutes a 'historical debt' social struggles owe to women and the relationship between society and nature, because these and other themes were often relegated by a vision that prioritized revolution. However, its practical execution is extremely complex.

In the case of the MST, the movement created a sector dedicated to gender-related issues, which besides creating a focused debate has also sought to discuss the relationship between patriarchy, *machismo*, socialization, and capitalism within and outside the movement. The autonomous actions of women and protests that emphasize the specificities and consequences agribusiness causes in the lives of women has been useful to show how it is impossible to dissociate the feminist struggle from the struggle for land and a decent life. Socialism and Marxism no longer appear as contradictory to environmentalism (a concern that for a long time was seen as a 'privilege of the rich'), and the MST found itself involved in the early stages of 'agroecological' transition, seeking to reorient the cooperativism of its settlements, incorporating into its economic plan the social and environmental dimensions of producing food free of contamination and pesticides and thereby challenge the agroindustrial exporting model.

I consider this incorporation of different thematic axes within the MST and other social movements' struggles to be a tendency towards *transversalization of collective action*, which, on one hand, forces social movements to generate new and increasingly complex frameworks and agendas, and, on the other hand, makes it incredibly difficult to create a typology of social movements based on thematic axes due to their multidimensionality and diversity.

AGGLOMERATING DEMANDS

As a consequence of the aforementioned process, the demands of the MST became increasingly complex. Although land concentration in Brazil is still alarming, the movement's struggle for agrarian reform is no longer restricted to the issue of land. It has become, in a broad sense, a struggle for territory that also involves the appropriation and management of natural resources, the participation and decisions on local and international commerce, sustainable development, the struggle for rights (to food, education, health, culture, information, etc.), and the dispute over production and development models.

As it gained national and international prominence in the last two cycles of contention, the MST incorporated into its agenda a series of struggles, demands, and interpretative frameworks. It also had to adapt to agrarian reform becoming market oriented and to the financialization of food and agriculture, which have culminated in a perverse process initiated in the first cycle of contestation with the international expansion of the agricultural market, its liberalization, and multinational corporations, which led to the commodification of food, the creation of an asymmetric and volatile agricultural structure, and ecological fragility (Clapp 2012).

The result of this process is the need to generate more complex demands that cover the multidimensionality of the contemporary agrarian issue. In the case of contemporary peasant movements, and of the MST in particular, all of these pieces have come together in its main contemporary claim: food sovereignty. This notion was created by the Via Campesina and appropriated by its member movements to supplement the limits of the struggle against hunger based on the notion of 'food security', described by most governments and the FAO as the physical, social, and economic access to enough food to satisfy the daily energetic needs. Although the notion of food sovereignty acknowledges the importance of adequate supply in order to satisfy necessities, for the MST and the Via Campesina it's not only a question of guaranteeing a sufficient quantity of food. We must also be concerned with the products being produced, how they are being produced, and the scale of production. Therefore food sovereignty is associated with the "right of peoples to define their own food and agricultural policies, protecting and regulating the national agricultural and animal production and commerce to meet sustainable development targets: determine to what degree they desire to be self-sufficient; restrict product dumping in their markets and give priority of use and rights over aquatic resources to communities that depend on fishing . . . food sovereignty does not prohibit commerce, it promotes the formulation of commercial practices and policies that respects the peoples' right to have an innocuous, healthy, and ecologically sustainable production" (Bringel 2012).

Food sovereignty is not restricted to the sphere of food. It encompasses a multidimensional array of themes, subjectivities, actions, and actors. That is why I suggest it is an *agglomerating demand*, in constant tension between the rural and the urban, the institutionalized and the institutionalizing, the protests and proposals, cooperation and conflict, networks and territories, the local and the global. Its appropriation by different social and political actors makes it a global demand, although one with different appropriations and types of local and national frameworks. This could be regarded as a tendency towards dispersion, although in practice it has been the common denominator that allows diverse social movements to understand each other and create common agendas, combining unity within diversity. They share a vision of issues that need to be tackled and broad demands, yet their resolutions have many specificities, which differs from the universalism of many global hunger and poverty policies.

FINAL WORDS

Social movements are not static. They rebuild themselves continuously according to their dynamics of struggle and social, political, and economic scenarios, reinventing themselves as they grow. In its almost thirty years of existence, the MST has redefined its allies and enemies and its agendas and demands, which are increasingly complex. The MST's struggle is not only a struggle for land and territory, but also a struggle for *multiple spatialities* (the places where it acts in the territories it fights for; the networks and different scales in which it participates; the dynamics of diffusion through which it exchanges experiences and creates common plans of action) throughout time. If the cycles of contention help to show the continuities and ruptures within the movement in relation to other actors and interactions between national politics and global contestation, it is only the self-reflecting potential and the capacity to adapt itself to the logics and the conjunctures of different cycles that make it possible for the MST to remain a model for peasant struggle in Brazil and the world. The fact that peasant movements, like the MST, have gone far beyond the rural struggles, localized and specific claims have challenged the very definition of what is being a peasant today. It deserves for future research a wider discussion on the restructuring of contemporary agrarian question, but also how contemporary peasant struggles within these new interpretive frameworks and social ties (including several of the elements discussed in this chapter) understand the boundaries of urban and rural, the local and the global, the specific and the universal. Maybe this would help us to illuminate the seemingly paradoxical question of why such localized rural struggles have acquired global dimension and why despite the decline in rural population in Brazil and other countries in the Global South there is an increasing visibility of the conflicts in the countryside.

NOTES

1. This chapter is part of a broader effort to analyze the internationalization of the MST and the role of this process in the reconfiguration of contemporary internationalism; the result of this effort is a book that will be published in 2013 in Brazil. This project is based on almost ten years of research on the MST and its internationalization, which took me to different locations in Brazil, Uruguay, Argentina, Venezuela, Guatemala, Mozambique, Senegal, Egypt, Portugal, Spain, France, Italy, the Netherlands, and England, for field

research and to compile bibliographical references as well as primary and secondary sources. I would like to thank members and former members of the MST's International Relations Secretariat (especially Janaina Stronzake, Geraldo Fontes, and Joaquin Piñeiro) for their interviews and conversations, which helped me to organize my observations.

2. One of the most emblematic examples is the cover of issue 2184 (September 2011) of *Isto É* magazine entitled "The End of the MST."

3. This classification into periods complements the one laid out by Fernandes (2000), the main reference for the historical reconstruction of the MST: genesis (1979 to 1985); territorialization and consolidation (1985 to 1990); territorialization and institutionalization (1990–). A more recent and complete analysis, focused not on the MST but on the relationship between social movements and the Brazilian state, Scherer-Warren (2012, ch. 3), also proposes a division into four 'phases,' although her division is based on decades: civic resistance movement against the authoritarian state (1960 and 1970), citizenship movement demanding the affirmation and regulation regularization of rights (1980 and 1990), institutionalized movement of participation and negotiation with the state (1990), and citizenship movement critical of social control through citizenship (2000s).

4. To learn more about each one of these organizations and the MST's perception of them, see the analyses of Santos de Morais (2006) and Morissawa (2001). For a broader examination see inter alia, the works of Martins (1981) and Medeiros (1989).

5. Religious commissions involved with rural workers.

6. Large and unproductive private farmland.

7. Some of these organizations include Christian Aid (UK) and Inter-Church Organization for Development Cooperation (Netherlands), Cáritas (Spain), and Frères des Hommes (France), created in the 1960s as a charity and converted in the 1980s to an NGO with no religious affiliation.

8. The MST National Congresses take place every five years and are an excellent 'thermometer' of the movement, because they are spaces for convergence and discussion, which analyze the past struggles, diagnose the present, and design an action agenda for the future. This is why I base my observations on the movement's reconstruction of its agenda mostly on the MST's congress. For the first four congresses, I utilize the movement's primary sources; for the fifth congress, which took place in 2007 in Brasília, I utilize my own field notes, as well my impressions as a participant observer.

9. The 2nd National Congress, which took place in May 1990 in Brasília, had more than 5,000 delegates and marked the movement's presence in nineteen states.

10. The state is seen as performing a dual function—an instrument of oppression and democratization, which is why it is considered an enemy by the MST, but also viewed as an important interlocutor for the formulation of public policies or other measures favorable to the movement.

REFERENCES

Borras, Saturnino, Jr. 2008. "La Vía Campesina and Its Global Campaign for Agrarian Reform." In *Transnational Agrarian Movements Confronting Globalization*, edited by Saturnino Borras Jr., Marc Edelman, and Cristóbal Kay, 91–121. Oxford: Willey-Blackwell.

Bringel, Breno. 2006. "El lugar también importa: Las diferentes relaciones entre Lula y el MST." *Revista NERA* 9: 27–48.

———. 2011. "A busca de uma nova agenda de pesquisa sobre os movimentos sociais e o confronto político: diálogos com Sidney Tarrow." *Politica & Sociedade* 10 : 51–73.

———. 2012. "Le Sans Terre du Brésil, l'activisme transnational et la souveraineté alimentaire comme alternative à la faim." *CERISCOPE (Sciences Po-Paris)* 2: 1–9.

Bringel, Breno, and Heriberto Cairo. 2010. "Articulaciones del Sur global: afinidad cultural, internacionalismo solidario e Iberoamérica en la globalización contrahegemónica." In *Descolonizar Europa descolonizar la modernidad: diálogos América Latina-Europa*, edited by Heriberto Cairo and Ramón Grosfoguel, 233–55. Madrid: IEPALA.

Bringel, Breno, and Enara Echart. 2010. "10 anos de Seattle, o movimento antiglobalização e a ação coletiva transnacional." *Revista Ciências Sociais UNISINOS* 48 (1): 28–36.

Brockett, Charles. 2005. *Political Movements and Violence in Central America.* Cambridge: Cambridge University Press.

Cardoso, Adalberto. 2010. "Uma utopia brasileira: Vargas e a construção do estado de bem-estar numa sociedade estruturalmente desigual." *Dados—Revista de Ciências Sociais* 53 (4): 775–819.

Clapp, Jennifer. 2012. *Food.* London: Polity Press.

Connolly, William. 2002. *Identity-Difference: Democratic Negotiations of Political Paradox.* Minneapolis: University of Minnesota Press.

Desmarais, Annette. 2007. *La Via Campesina: Globalization and the Power of Peasants.* London: Pluto Press.

Fernandes, Bernardo Mançano. 2000. *A formação do MST no Brasil.* Petrópolis: Vozes.

Hellmann, Michaela. 1995. "Democratização e movimentos sociais no Brasil." In *Movimentos sociais e democracia no Brasil*, organized by Michaela Hellmann, 9–23. São Paulo: Marco Zero.

Kowarick, Lucio. 1987. "Movimentos urbanos no Brasil contemporâneo: uma análise da literatura." *Revista Brasileira de Ciências Sociais* 1 (3): 38–50.

Martins, José de Souza. 1981. *Os camponeses e a política no Brasil: as lutas sociais no campo e seu lugar no processo político.* Petrópolis: Vozes.

Medeiros, Leonilde. 1989. *História dos movimentos sociais no campo.* Rio de Janeiro: FASE.

Morissawa, Mitsue. 2001. *A História da luta pela terra e o MST.* São Paulo: Editora Expressão Popular.

Olesen, Thomas. 2004. *International Zapatismo: The Construction of Solidarity in the Age of Globalization.* London: Zed Books.

Pleyers, Geoffrey. 2010. *Alter-Globalization: Becoming Actors in a Global Age.* Cambridge: Polity Press.

Rovira, Guiomar. 2009. *Zapatistas sin fronteras: Las redes de solidaridad con Chiapas y el altermundismo.* México D.F.: Ediciones Era.

Sader, Eder. 1988. *Quando novos personagens entraram em cena: Experiências e lutas dos trabalhadores da Grande São Paulo: 1970–1980.* Rio de Janeiro: Paz e Terra.

Santos de Morais, Clodomir. 2006. "História das ligas camponesas do Brasil." In *História e natureza das Ligas Camponesas: 1954–1964*, edited by João Pedro Stédile, 21–76. São Paulo: Expressão Popular.

Scherer-Warren, Ilse. 2012. *Redes emancipatórias: nas lutas contra a exclusão e por direitos humanos.* Curitiba: Editora Appris.

120   *Breno Bringel*

Scherer-Warren, Ilse, and Paulo Krischke, eds. 1987. *Uma revolução no quotidiano?: Os novos movimentos sociais na América do Sul.* São Paulo: Editora Brasiliense.
Slater, David. 2000. "Repensando as espacialidades dos movimentos sociais. Questões de fronteiras, cultura e política em tempos globais." In *Cultura e política nos movimentos sociais latino-americanos,* edited by Sônia Álvarez, Evelina Dagnino, and Artro Escobar, 503–33. Belo Horizonte: Editora UFMG.
Tarrow, Sidney. 1998. *Power in Movement: Social Movements and Contentious Politics.* Cambridge: Cambridge University Press.
———. 2012. *Strangers at the Gates.* Cambridge: Cambridge University Press.
Vieira, Flávia Braga. 2011. *Dos proletários unidos à globalização da esperança: Um estudo sobre internacionalismo e a Via Campesina.* São Paulo: Alameda.

7 Brazilian Culture as Category of Public Intervention

Myrian Sepúlveda dos Santos

INTRODUCTION

Anderson's *Imagined Communities* (1983) disseminated the idea that different nationalities were built as cultural artifacts sharing in common the capacity to command profound emotional feelings among nation's citizens. Just how were they built and in what ways they have changed their meanings have become important issues.

The building of national identity will be considered along with global, national, and local political, social, and economic dimensions. Yet the interdependence between nations does not mean the same processes are at work. Certainly, cultural policies are related to class struggle, social movements, elite's interests, hierarchical structures, and inherited traditions. Although these are all crucial elements to the understanding of the construction of national identities, I do not believe that there is a single model, factor, theory, or pattern capable of explaining how they intertwine with one another and operate. Struggles for power and wealth have different actors, structures, and histories in each country. Although some common trends may be found in the ways in which national identities have been shaped, these identities are intertwined with the formation of the nation-states themselves, making them far-reaching processes that can be better understood in light of each nation's distinct historical formation.

The choice to focus on public policies is related with the strong role of the state in building the image of the Brazilian nation. Despite periods of democracy in which citizenship and social participation have expanded, Brazilian republican life can be characterized by a centralized and strong state power in the hands of a small elite. The formation of the nation began with the arrival of the Portuguese court and its administrative structure in 1808. From the earlier colonial and slavery system to the nineteenth-century empire and the subsequent republican regime, Brazil has presented high levels of inequality and extreme poverty. As a peripheral nation under modern capitalism, the economy was based on the exports from agricultural, mining, and manufacturing sectors. This economic system allowed the concentration of power and wealth. Along with social inequality, most of the internal workforce was informally employed, leading to an absence of legal protection for the majority of the economically active population. The political system was constituted by clientelist and paternalist relations, that is to say, by the exchange of votes for small favors and by the maintenance of personal ties. Despite periods of democratic norms and rules, the economic and political systems of the nation guaranteed the perpetuation of power in the hands of its elite.

In the twentieth century, a very strong Brazilian elite of wealthy landowners and industrialists was joined by middle sectors of the population. The latter consisted largely of skilled workers and public servants whose power was based more on knowledge than property. However, the great majority of the population remained excluded from education and formal jobs. Brazil was a nation divided into full citizens and a poor population excluded from the public space. Poverty was entangled with ethnicity. Freed slaves and their descendants, people of indigenous origin, and those who are identified by *mestiçagem* were treated as delinquents, lazy, and incapable people, stigmas that persist today. Poverty reached its highest levels in rural areas of the Northeast and the peripheries of the major cities. As is frequently observed, some areas of Brazil look like Southern California with their luxury high-rise condominiums, whereas others are more reminiscent of the poor communities of Haiti.

Considering these high levels of inequality in Brazilian society, Edward Said's question is crucial to understanding the building of national identities: what happens when a nation or a nation's segment assumes to know more about you than you do about yourself? (Said 1978, 1985). Brazil's national identities are constructions that have conferred positive attributes to those members of the nation holding power and negative ones to those who do not. Powerful social groups have created narratives that legitimate their positions, habits, and desires. These self-conferred positive attributes have coexisted with the stigmatizations projected onto other groups. Both sets of values have been reproduced by social actors and appear like natural qualities.

Processes of national identity building involve power. The power of narratives and cultural practices in shaping people's lives, habits, and desires has been stressed by various key authors. As indicated, Said's *Orientalism* explored an important issue by denouncing the political use by Western nations of 'Orientalism' as a constellation of ideas and practices. As is well known, his theses were influenced by concepts such as Gramsci's cultural hegemony and Foucault's analyses of the power of knowledge. Unlike Foucault, Said believed that citizens could change their own destinies. And unlike Gramsci, Said invited a particular analysis of the intertwinement of culture and politics that was not determined by class struggle. Said's work has been highly influential in postcolonial studies. After Said's work, important questions were raised by all those taking culture and power seriously: Who represents whom in collective constructions? What are the consequences of these representations? Therefore, it is not just a question of 'how' but also of 'who'.

As a matter of fact, the monopolization of cultural resources combined with the exclusion and stigmatization of members of other groups was not in itself an innovation, because these aspects are found in many theoretical approaches. The struggle for ideological and political hegemony transcends political boundaries. The main challenge to those theories is to acknowledge the constraints within narratives without denying people's capacity to build their own lives and create their own representations. For Said, as Orientals were represented by others and were unable to impose their own view about themselves, their identities became marked by a hybrid or dual character. In Brazil, as will be shown, although subaltern groups lack the power to represent themselves to a larger community, they have the capacity to build a representation of themselves and are able to contest the dominant representation. Cultural and social manifestations experienced as apolitical represent different ways different groups resist oppression.

THE BEGINNING: A NATIONALIST-POPULIST POLICY

In 1929, the United States was going through one of its worst economic crises with global repercussions. In Brazil the large São Paulo coffee exporters also entered into crisis and with them the oligarchic governments that had dominated the republic. The extreme concentration of land and power seemed to be defeated. Getúlio Vargas took power in 1930 based on a new social pact. He favored industrialization, created state monopolies for strategic sectors, and promoted policies of social inclusion. In 1937 Vargas closed Congress, banned public demonstrations, and imposed strict censorship on the media, installing an authoritarian government that eroded notions of citizenship. Support for his government depended fundamentally on corporatist policies that gave rights to urban workers, diluting any perception of class conflict. Along with these proposals, the interventions in the cultural sphere were essential to the process of legitimizing his government. Brazilian national identity achieved a stable configuration during the Vargas dictatorship and was maintained with little variations until the late 1990s by liberal and authoritarian regimes alike.

At the beginning of the 1930s, Brazilian intellectuals challenged European racial theories as they valued the idea of a nation formed by the amalgamation of three races: white, Indian, and black. Gilberto Freyre, one of the prominent intellectuals of his time, helped enhance the standing of the *mestiços*, people of 'mixed blood', described as inferior by the racial theories of the time. Moreover, he placed a positive spin on the relations between the 'manor house' and 'slave quarters', and strengthened the belief in racial democracy. As Ortiz (1991) points out, the *mestiço* became national. Notwithstanding, the union of the races became accepted only within the wider conceptual framework of the gradual whitening of the Brazilian population, an ideology that presumed the superiority of the white race and led to practices of racial discrimination that proved difficult to combat.[1]

Vargas's cultural policy was responsible for the construction of a nationalist-populist imaginary that remained dominant until the 1990s. Under the direction of Gustavo Capanema, minister of education and health, an artistic and cultural avant-garde linked to the modernists began to contribute with the government project. The idea of racial democracy was incorporated as one of the nation's virtue. The official reinvention of Brazil was pursued in a nationalistic and exacerbated form, emphasizing the exceptionality of the Brazilian people, its heroes, the exuberance of its natural environments, and its technical and scientific development. In 1937, a few days after the coup that led to the 'Estado Novo' (New State), the authoritarian government created several institutions and laws in the cultural sphere.[2] The National Historical and Artistic Heritage Service (Serviço do Patrimônio Histórico e Artístico Nacional [SPHAN]), directed by Rodrigo Mello Franco de Andrade between 1937 and 1967, became one of the country's most important and stable cultural institutions.[3] The institution invested in the conservation of monuments that embodied the nation's elites and the state administration. Besides modernist architecture, constructions deemed worthy of preservation included colonial-style buildings such as forts, old city halls, manors, and mansions. The baroque legacy contained in religious art was exalted as the nation's truest form of artistic expression.

The original proposal for creating SPHAN, drafted by one of the leaders of the modernist movement, Mario de Andrade, included an interest in preserving the cultural practices of the 'popular masses'.[4] 'Popular' culture, however, was largely ignored by preservationist practices; they were useful only as political instruments; as living culture, they formed the core of a new policy in relation to the masses. Bans were lifted on practices such as samba and capoeira, and some of their leaders gained official support. Institutions were created to repress themes that could conflict with patriotic, historic, and educational themes. The culture industry was seen as an important tool in economics and politics. Protectionist laws were passed to shield national production in book publishing and in the record and film industries, as well as provide incentives in the dramatic arts. Primarily through radio programs, cultural manifestations such as samba and carnival became national emblems. A cultural split was formed between preservation practices and museums, on one hand, and popular manifestations, on the other hand.

After the Second World War, Brazil entered into a democratic period.[5] In the economic level, the domestic market was opened up to large foreign companies, and the import substitution policy led to a growth in the economy and industrial production. For the first time the industrial sector overtook the agricultural sector. This development policy was accompanied by an unprecedented urbanization, an intense flow of migrants from rural areas to the cities, large-scale social inequalities, and unacceptable levels of poverty. The largest portions of the population continued to be absorbed into wage labor in precarious forms.

In the area of culture, private initiatives replaced government investment, looking for the urban middle sectors of the population, which enlarged the internal consumer sector. The influence of the American lifestyle, exported by the media, led to an increase in the consumption of culture and leisure activities.[6] Public initiatives were replaced by private capital in cinema, theater, and music. Bossa nova, a new musical style that combined samba with jazz-like harmonies, produced by middle-class musicians in Rio, became famous nationally and internationally.

Despite the various criticisms concerning the artificiality of the division made between folklore and popular culture, the National Folklore Commission was created in 1947 with connections to UNESCO (the United Nations Educational, Scientific and Cultural Organization). This organization was created in 1945 and had a cultural agenda to protect indigenous cultures and cultural traditions threatened by the rapid economic development experienced by industrialized nations, presupposing a basic antagonism between the two poles. In 1953, the Folklore Defense Campaign, the first permanent Brazilian institution dedicated to the field, was created.

THE LEFT-WING POPULAR NATIONALISM AND ITS DECLINE

The revival of the nationalist project grew in strength in the 1960s, this time led by left-wing intellectuals and artists. The harmonic idea of a homogeneous nation based on the union of races became contested. The questions of national identity and culture industry divided an entire generation. On one hand, it was believed, within the classic framework of the Frankfurt School's critique, that the rules of the market and the mass production of cultural products made it impossible for artists to transmit their experience of art to the public.[7] On the other hand, criticism of the former nationalist-populist ideology grew along with the search for new paths to incorporate the great 'masses'. Artists and intellectuals attempted to give a new meaning to national identity.

In the 1960s, Brazilian organized movements, involving intellectuals, artists, and students, strongly influenced by the left and an anti-American feeling, demanded a more equitable distribution of income and agrarian reform. These movements became strong among middle sectors of the population. It was the golden age of engaged art when intellectuals took any kind of dissidence as alienated and inconsequential conformism.[8] The African American civil rights movement in the United States also influenced Brazilian social movements. In the 1960s, therefore, the country's leading artists and intellectuals were involved in the production of an aesthetic that opposed the logic of the market.[9] Inspired by popular culture movements in France,[10] intellectuals from the Brazilian Northeast, among them Paulo Freire, sought out new methods of raising the population's awareness through folk theater and grassroots literacy and educational campaigns. Universities created the famous CPCs, Centers for Popular Culture, which divulged art through critical and politicized texts.

The 1964 military coup repressed popular, black power, and left-wing movements and subjected the nation to authoritarian rules. In the mid-1960s, television was already the most important medium of communication. The military government opened Brazil to imported programs and saw the new media as a potential tool for creating a capitalist society. Despite the new government's investments in the cultural industry and the spreading of the North American way of life, left-wing intellectuals continued to be critical of Western capitalism and modern technical society. Musical festivals transmitted by local television networks attracted the young and became highly politicized. They were important social events, because they represented public spaces within an authoritarian regime where different political ideas about autonomy and nationality struggled with each other.[11]

May 1968 is mostly known by the outbreak of the general strikes in France and a series of protests in several capitalist countries. To members of the French Communist Party (PCF), these movements were not revolutionary, but essentially demands for economic reforms. As we know, social movements do not go only in one direction. In Brazil, the widespread student protests raised a brutal repression. The government declared a nationwide state of siege, closed the National Congress, suspended habeas corpus, and adopted brutal measures such as political assassinations and torture. During this period, the economic success was associated with higher levels of inequality. Whereas the Brazilian Gini coefficient was around 0.50 in the 1960s, in the following decades the level of inequality was even higher: 0.62 in 1977 and 0.58 in 1986.[12] Unable to mobilize the working class and even the poorest sectors of the population, left-wing groups adopted guerrilla tactics.

Paralleling these repressive measures, the military dictatorship associated nationalism with the defense of the union against communism. In the 1970s, extensive censorship coexisted with noncritical cultural productions made for mass consumption: encyclopedias, imported musicals, CDs, sexploitation films, and *telenovelas*. The government invested in new technology and integrated the country by building national microwave and satellite distribution systems. Television sets were subsidized and pushed into the population. Without any substantial project for strengthening public TV, the government promoted the growth of one private network, Globo Organizations, allowing the creation of a monopoly in the mass media. Publishing gained an enormous boost with the policy of stimulating the paper industry.[13]

The military regime remained the main investor in practices associated with heritage preservation. In 1973, the Brazilian government invested a large amount of funds in the Historical Cities Reconstruction Program (PCH). The homogeneous and unified nation continued to be based on the same emblems raised in the 1930s: the racial democracy ideology and the elite's cultural legacy. The 'popular', however, was associated with the market. The visual artist and designer Aloísio Magalhães, who became director of SPHAN-FNPM,[14] developed a new interpretation of popular crafts, searching for a hallmark of 'Brazilianness' that could be stamped on cultural goods. Besides that, in 1968 the Museum of Folklore was created in Rio de Janeiro, serving as the head office for the Campaign for the Defense of Brazilian Folklore. In 1983 the Room of the Popular Artist (SAP) was created in the grounds of Rio's Museum of Folklore with the aim of selling craftwork made by renowned artisans.

OPENING TO CULTURAL DIVERSITY

The expansion of democratic values reached Latin America nations in the late 1970s. In 1979 the Brazilian amnesty law was approved after a series of large-scale popular demonstrations. The break with the authoritarian state was consolidated with the promulgation of the 1988 Constitution, which limited the power of the state and guaranteed individual, political, and social rights. In the most industrialized region of the country (São Paulo's ABC), labor unions held strikes and resisted government pressure. Along with direct, secret, and universal voting, the Constitution legitimized participation mechanisms and cultural diversity. It became a crime punishable by law to commit any act of discrimination based on race, color, ethnicity, religion, or nationality. Created during the same year was the Palmares Cultural Foundation, a public institution linked to the Ministry of Culture. This was the first official institution dedicated to fighting racism and strengthening the preservation, protection, and dissemination of the culture of Afro-descendants.

Despite the democratization process, the concentration of wealth and land continued to be the rule. The Gini coefficient was 0.602 in 1997. High levels of inequality represented lack of participation, citizenship, and rights to the majority of the population. Although to a less intense degree than other Latin American countries like Chile and Argentina, Brazil also adopted neoliberal measures such as greater opening to international trade, privatization of public companies, and flexibilization of the labor market. The entry of new technologies, which drastically reduced the number of workers in the industrial sector, combined with labor insecurity weakened the union movements. Already a feature of European and North American democracies, new social movements boosted the fight for recognition of causes and rights in Brazil.[15] In the rural world, the Landless Rural Workers Movement (Movimento dos Trabalhadores Rurais Sem-Terra [MST]) grew in strength.

In 1985, tax incentives were introduced to allow companies to invest in culture. Due to the revenue obtained by the country's largest companies, they ended up concentrating the decision-making power over cultural investment, supporting already established artists and cultural productions capable of attracting a large public and avoiding nonprofit cultural manifestations.[16] The only area in which the government continued to invest directly was in heritage, an intervention connected with international policies. In 2000, the government created the National Intangible Cultural Heritage Program, ratifying UNESCO's recommendations[17] concerning the necessity to preserve traditional and popular culture, considered part of humanity's universal heritage. The measures continued to be justified by the imminent danger of 'traditional and popular' culture disappearing in the face of industrial and technological development.

Despite the economic and political changes and even the upsurge of new social movements, the government reinforced the model of unified nation constructed back in the 1930s. In April 2000 the country's largest ever exhibition opened with the support of businesses and the government. A total success with the public in Brazil and beyond, the *Brazil 500—Rediscovery Exhibition* celebrated 500 years of the discovery of Brazil. The different rooms of the exhibition displayed rock paintings and various forms of indigenous, Afro-Brazilian, popular, modern, and contemporary art. There was no questioning of the use of the concept of a work of art in indigenous nations, or any critique of Portuguese colonialism, the massacre of indigenous populations, or even the immense tragedy of slavery.

THE MULTICULTURAL NATION

The election of President Luiz Inácio da Silva (Lula) in 2003 was achieved through a coalition of center-left parties. The government maintained several elements of the former liberal economic policy and consolidated the exportation of commodities. Despite following the liberal economic agenda of previous governments, the new president gave priority to income distribution and promoted a social reform focused on the poorest. Goods and resources were distributed through a series of interconnected programs and resulted in a reduction in poverty and social inequalities, incorporating new sectors of the population in the market and increasing consumption.[18] Although Brazil has continued as one of the countries with the highest income concentrations in the world, the Gini coefficient diminished from 0.58 to 0.54, between 2003 and 2009. The income of poor Brazilians rose seven times faster than the income of the rich for the first time in the nation's history. Paralleling the fall in poverty there was an increase in democratic rights. Three special secretariats, directly linked to the presidency of the republic, were created: the Special Secretariat for Policies for the Promotion of Racial Equality (SEPPIR), the Special Secretariat of Policies for Women (SEPM), and the National Youth Secretariat. The Special Secretariat for Human Rights (SEDH), created by the previous government, was also maintained. Social movements, trade unions, black, indigenous, landless, and women activism all acquired certain rights that were previously denied. These special secretariats contributed to the replacement of the 'national-popular' model by one emphasizing Brazil's cultural diversity. All these changes were related to new policies on the culture area.

The cultural patronage laws responsible for most of the private investments in the culture area were maintained. But different from the previous liberal governments, the new government directly intervened in the culture area with the creation of new institutions, policies, and laws, and higher public funds. [19] The musician and composer Gilberto Gil, minister of culture between 2003 and 2008, was responsible for national policies directed to both social inclusion of the poorest and protection of diversity. The National Culture System (SNC) was created to connect and integrate the diverse cultural policies; the National Culture Plan (PNC) established public policies; and the Constitutional Amendment no. 48, of August 10, 2005, ratified the PNC and set the targets of access to cultural goods and the valorization of ethnic and regional diversity. Official data from 2008 showed that 90 percent of the Brazilian population did not go to theaters, cinemas, libraries, and bookshops. [20] Setting out from the premise that every citizen has a right to culture, the government worked to improve the access of low-income populations to existing activities. Moreover, specific programs, led by the 'Culture Point', were created to promote artists coming from small communities and regions far from the metropolitan centers. In these cases, the government distributed financial and technological resources to a varied number of local projects. Policies were introduced to integrate the theme of diversity with the promotion of citizenship, reinforcing concepts that for the most part the United Nations had been pursuing at the international level since the 1980s, and that had been embodied in the 1988 Brazilian Constitution. [21]

Although culture industries tend to predominate in the culture area, they are not incompatible with the construction of values and sensibilities. [22] At the end of the 1960s, Gilberto Gil was one of the leading figures of the 'Tropicália', a countercultural musical movement that challenged the authoritarianism embedded in the formulation of the 'national-popular'. In power forty years later, the artist gave priority to cultural diversity. His actions also defied market laws as he defended new digital technologies, the free circulation of culture, the relaxation of copyright laws, and the distribution of cultural works through Creative Commons licenses. New policies supported nonprofit making activities like educational films, circus acts, classical music, and public television. Cultural activities in small towns and working-class districts also received public support.

The government intervention in the heritage area has proved to be more problematic. For the first time, monumental buildings and folk culture were placed under the administration of a single institution, IPHAN (the National Institute of Historical and Artistic Heritage). The institution responsible for safeguarding the country's heritage focused its attention on the registration of intangible cultural heritage, which became a state policy. Immaterial or intangible culture—that is, orality, forms of knowledge, rituals, and everyday practices—was conceived an alternative to the classical canons of what was considered to be the Eurocentric heritage. Following the UNESCO Convention for the Safeguarding of the Intangible Cultural Heritage, created in 2003, the government understood that the official protection of the intangible culture could favor those countries that entered the process of industrialization at later dates, as well as sectors of the population thus far ignored as producers of culture.

In the attempt to democratize the idea of culture, the government adopted the idea of the 'anthropological concept of culture', a term that became synonymous with meanings and values inherent to any human interference in the world. By generalizing the concept of 'culture', the Ministry of Culture's target became any cultural manifestation. The new definition of culture undermined any selective criteria to preserve popular arts.[23] In Brazil, acarajé sellers, capoeira masters,[24] and other social groups that had been previously confined to the informality of the market began to demand labor rights from the moment that their activities were legitimized by the state. These are rights, however, that have demanded the participation of organized groups and intellectuals. These latter have had the power to point out which manifestations should be preserved. It must be added that the registration of intangible cultural heritage standardizes artisanal and informal practices, diminishing hybrid constructions and accelerating their process of adaptation to the consumer market.[25] Both tangible and intangible heritage are important components of the global tourist industry and of the subsequent phenomenon of 'gentrification' in the processes of revitalizing urban centers.[26] In addition, as we consider the United Nations' policies, it is possible to observe that the UN commissions do not value tangible and intangible culture in the same way; there is a hierarchy between nations, their cultural heritage, and markets.

CONCLUSION

In Brazil we have seen that between 1930 and the end of the 1990s, while liberal governments looked to avoid a direct intervention in the cultural arena, the dictatorships ensured that public administration worked to implement strong cultural policies. However, all these governments, without exception, reproduced and strengthened a unified construction of the nation originally implanted at the start of the 1930s. Over the last decade, the PT government has revived the interventionist tradition, but replaced the unified image of the nation with approaches that prioritize diversity and difference. The weakening of the nationalist-populist project enabled the strengthening of large social groups oppressed by centuries of colonization. Affirmative action programs for African-descendant, women, indigenous peoples, and other social groups have been encouraged by the government. The belief that a unified subject could represent the different social demands fell apart.

Therefore, in a Brazilian multicultural nation, marginalized communities fight for the right to 'speak for themselves' after centuries of oppression. In a 2007 study, the IBGE showed that 'whites' receive and average monthly income almost twice that of blacks and pardos. The 2010 IBGE Census also showed that whereas 23 percent of 'whites' have more than 12 school years, the percentage for the same number of school years of those who are identified as 'blacks', 'nigers', 'morenos', or 'pardos' varies from 4.3 to 11.8.[27] The indigenous peoples comprise a large number of distinct ethnic groups. Their population has declined from a pre-Columbian high of millions to 500,000 in recent times, what can be considered one of the largest genocides in human history.

The intention of this chapter is to highlight the entanglement between official policies on the culture area, the building of the national identity, and citizenship. Federal policies and affirmative action have improved the self-esteem of different social groups and the distribution of public resources. As demands for actions against race discrimination, for instance, achieved governmental support, Brazilian former perceptions of racial democracy have been changing faster. The proportion of Brazilians who self-report to be Afro-descendant is growing faster, and the last IBGE Census (2010) showed that for the first time the majority of the population (50.7 percent) considered themselves 'black' or 'pardo'.

As asserted by Nederveen Pieterse (2007: 102) there is no monolithic approach to multiculturalism. In the Brazilian case, the multicultural way of building the national imaginary emphasized populations that predated the formation of the nation-state in the nineteenth century. The same author calls attention to the different ways collective and individual rights interrelate (2007: 123). As a matter of fact, what is at stake is whether Brazilian campaigns for recognition and redistribution run side by side without necessarily competing.[28] As an example, we can take the demands for the right to land. MST (the Landless Rural Workers Movement) is the strongest Brazilian social movement fighting for the right to land. In August 2000, MST held its Fourth National Congress in Brasília, campaigning for agrarian reforms. The modernization of the agro-export economy in the 1960s and 1970s intensified the concentration of land and wealth, provoking a rural exodus and sustaining older authoritarian power structures. Rural workers were forced to move to areas with low soil productivity, meaning that more than half of the millions of Brazilians living below the poverty line reside in rural areas. As the leading sector of the Brazilian economy, the agro-export model inserted the country in the international market on the basis of an increased exploitation of rural workers. As was already said

Brazil continues to be one of the nations with the higher levels of social inequality in the world. The land concentration continues to be guaranteed by federal laws as well as by means of brutal and ignominious violence. The MST members fight for their own properties and against the persistence of the concentration of land and high levels of inequality.

After the 1988 Federal Constitution, two other important groups became noticeable in the public sphere because of their fight for the right to land: indigenous and Afro-descendant peoples. In Brazil the concentration of lands began with the beginning of the European colonization and the consequent appropriation of everything contained in them. In 1988, after years of nondemocratic systems, the Federal Constitution gave special emphasis to human rights and recognized the territorial rights of indigenous peoples and *quilombolas* (descendants of slaves who lived in *quilombos*) based on their traditional heritage. During the colonial period, the *quilombos* were places of refuge for fugitive slaves, generally situated in fairly inaccessible regions. With the recent promotion of diversity, the nation's history of oppression ceased to have a single reference point. Indigenous peoples and Afro-descendants have rewritten the history of the nation and have mobilized cultural criteria of self-recognition in order to fight against all sorts of oppression.

Institutions such as the Palmares Cultural Foundation were created to fight against racial discrimination. The institution has already recognized more than 3,500 rural and also urban *quilombos*. Nonetheless another federal agency, the National Institute of Colonization and Agrarian Reform (INCRA), which issues the land title, requires several documents such as certified proofs of the authenticity of the *quilombo* by specialists from various areas. In 2004 the government launched the Quilombola Brazil Program to reinforce the Federal Constitution. As historical records of former *quilombos* have been lost in time, the criterion of self-identification, already used by Afro-descendants for access to universities through quota systems, was adopted. Yet few titles were issued to this day.

In the case of the indigenous peoples, the attempt to save them from exploitation and decimation is older and less efficient. Brazilian government established the National Service for Protection of the Indians (Serviço Nacional de Proteção aos Índios) in 1910, an institution that was replaced in 1967 by the National Indian Foundation (Fundação Nacional do Indio [FUNAI]), a governmental council on Indigenous issues. The struggle on behalf of indigenous peoples involves the defense of their rights over lands and natural resources. Brazil has more than 200 small indigenous nations and they require the demarcation of land with their resources protected. The democratization process of the last two decades has not guaranteed their rights. Brazilian governments have assured neither the demarcation of indigenous peoples' territories nor the protection of natural resources.

The conflict of interests extends to clashes between the policies implemented by different government bodies. Before emitting any certificate of ownership, the INCRA, a powerful institution that did not change their bureaucratic system, requires a huge number of documents, most of them hardly available for poor communities. The processes are slow and those companies that invest in the exportation of commodities, energy production, or even in the Amazon rain forest for mining, logging, farming, and cattle ranching usually are the winners. Violent confrontations and all kind of brutalities continue to be the rule.

There is a growing defense of fixed traditions by *quilombola* communities and indigenous peoples who, along with the recognition of their identities, fight for rights previously denied to them. The federal policies based on diversity strengthen group boundaries. Those who fight for land as members of the MST are poor people who certainly have as ancestor the same indigenous and Afro-descendant peoples. The identities mobilized by different groups are not based on any irrational belief that their current practices are a direct continuation of those from time immemorial. In many cases the claim for essential roots is the result of a conscious rhetoric or strategy to achieve resources and official support.[29] However, essentialism allows those who defend tradition to enter into conflict with one another in awkward situations, because the hybridism constituted between ethnic groups and cultures is spread in rural areas. Recent studies show that members of a single community break apart in order to perform different identity roles in their strategy for obtaining reparation and social visibility.[30]

As we have seen, after 2003 and the arrival of the Lula government backed by a center-left coalition, the homogeneous and unified image of the Brazilian population was replaced by an idea of a nation composed of cultural differences that attempted to do justice to the complexity of its demographic layers. The nation opened its doors to historical reparation. However, it has not been easy for either indigenous peoples or *quilombolas* communities to obtain justice in the present for experiences of oppression inherited from the past. In terms of land rights, few have succeeded. Their vulnerable social conditions result present inequities as much as past ones. It is undeniably the case, therefore, that whereas some government sectors are concerned with 'who speaks', those holding power in the lands distribution benefit from the political fragmentation of social actors and groups unable to build a coalition politics. The political balance still favors powerful entrepreneurs and those advocating developmentalism as national policy. Yet the outcome of these conflicts is never definitive, and although building a coalition politics capable of bridging differences between landless workers, *quilombolas* communities, indigenous peoples, and other rural and subaltern groups may appear extremely difficult, the possibility always remains that it could succeed.

NOTES

1. On Freyre, see especially *Casa Grande e Senzala* (Freyre 1933). For an excellent analysis of eugenic practices, see Stepan (1991).
2. For a more detailed description of the process of institutionalization in the culture area during the New State, consult Miceli (1984, 2001) and Williams (2001).

3. Created in 1937, SPHAN went through various alterations to its administrative structure over the years. For a deeper analysis of the process of preserving Brazil's heritage in the 1930s and 40s, consult Chuva (2009).
4. On Mario de Andrade's proposal, see Chagas (2006).
5. On populism in Brazil, see among others Ianni (1968), Laclau (1977), Weffort (1978), and Gomes (1994).
6. See, for instance, the creation of art galleries, exhibitions, and museums such as the São Paulo Museum of Art (MASP), the São Paulo Museum of Modern Art, the Rio de Janeiro Museum of Modern Art, and the São Paulo International Biennale.
7. On this topic, see Adorno and Horkheimer (1997).
8. For a range of perspectives on the debate surrounding art, politics, and nationalism during this period, see among others Schwarz (1979), Buarque de Hollanda (1981), Ortiz (1986), and Ridenti (2000).
9. See, for instance: Helio Oiticica was renewing visual language through his neo-concrete art, creating his famous *parangolés*, Glauber Rocha gained international prestige in Cannes with the new cinema, and José Celso Martinez Correa brought the modernist poet Oswald de Andrade to the theater.
10. On this point, consult Dubois (1999).
11. For more detailed information on the 1967 Festival competition shown by TV Tupi, the most emblematic of the period, see the film *Uma noite em 1967*, directed by Renato Terra and Ricardo Calil. Young musicians such as Edu Lobo and Chico Buarque, working closely with Brazilian musical styles like samba and northeastern music, became the representatives of a musical trend later called MPB (*música popular brasileira*: Brazilian popular music). Brazilian rock or *iê-iê-iê* was represented by Roberto Carlos, Erasmo Carlos, and Vanderleia, also called the Jovem Guarda (Young Guard). A small group of musicians, consecrated in 1968 with the LP Tropicália, questioned the constraints imposed by the idea of 'nationalism' whether built from left- or right-wing policies. There is a growing number of analyses, articles, and books on the tropicalista movement, especially among North American authors. On this latter subject, see the book by Caetano Veloso himself (1997), but also Buarque de Hollanda (1981), Perrone (1989), Sovik (2001), and Dunn (2001).
12. For a historical series of the Brazilian Gini coefficient, a statistical measure of social inequalities, see Instituto de Pesquisa Econômica Aplicada (IPEA) (n.d.).
13. For analyses of cultural policies and the development of the culture industry during the period, see Arruda (2004), Miceli (1984, 2001), and Ortiz (1986, 1991).
14. Created in 1979, this structure combined the National Secretariat Historical and Artistic Heritage (SPHAN), a body responsible for directing and coordinating policy at the national level, and the National Pro-Memory Foundation (FNPM), the agency responsible for providing the means and resources for implementing the activities, was maintained until 1990.
15. On the emergence of these new movements, see Escobar and Alvarez (1992).
16. For comparative data on cultural projects supported by national funds, see Ministério da Cultura (n.d.).
17. See, for instance, Recommendation 1989 (Traditional Culture and Folklore); Universal Declaration 2001 (Cultural Diversity); Universal Declaration 2003 (against the deliberate destruction of Cultural Heritage); Convention 2003 (Intangible Cultural Heritage); Convention 2005 (Diversity of Cultural Expressions).
18. During this period we can cite the increase in the minimum wage, the curbing of child labor, the 'Family Allowance' program of direct income transfer

to impoverished families, and 'Hunger Zero', a program directed towards nutrition and food security.
19. Government investments in the culture area rose from R$111.6 million in 2003 to R$550.6 million in 2009. Total patronage in 2009 was around R$1.100 million. See Ministério da Cultura (n.d.).
20. For data on cultural access, see Instituto Brasileiro de Geografia e Estatísticas (IBGE) (Brazilian Institute of Geography and Statistics), PNAD (Pesquisa Nacional por Amostra de Domicílios) (2008).
21. See, for example, the UN World Conferences on human rights abuses (Vienna 1993), children (New York 1990), victims of racism (Durban 2001), and women (Beijing 1995), as well as those on transnational themes such as the environment (Rio de Janeiro 1992), population (Cairo 1994), social development (Copenhagen 1995), and housing (Istanbul 1996).
22. On the relation between citizenship and the market, see García Canclini (1995).
23. On the new concept of culture, see, for example, Ministério da Cultura (2010).
24. Acarajé is a savoury Bahian dish made from deep-fried balls of mashed beans, served with a spicy paste, salad, and shrimp; and capoeira is an Afro-Brazilian dance and music that like other martial arts incorporates self-defence maneuvers.
25. On themes of cultural hybridism in Latin America and the commercialization of heritage, see García Canclini (1990).
26. For an analysis of the relation between world heritage and global tourism, see Peixoto (2000).
27. See Instituto Brasileiro de Geografia e Estatística (IBGE) (n.d., 2007).
28. The compatibility between politics for redistribution and recognition has been the issue of huge controversies. On this issue, see Fraser (1998) and Young (1998).
29. See Castro (2006).
30. On this topic see, among others, Arruti (2005) and French (2009).

REFERENCES

Adorno, Theodor W., and Max Horkheimer. 1997. *Dialectic of Enlightenment.* London: Blackwell.
Anderson, Benedict. 1983. *Imagined Communities.* London: Verso.
Arruda, Maria Arminda. 2004. *A embalagem do sistema: a publicidade no capitalismo brasileiro.* 2nd ed. São Paulo: EDUSC.
Arruti, José Maurício. 2006. *Mocambo: Antropologia e História do processo de formação quilombola.* São Paulo: EDUSC.
Buarque de Hollanda, Heloisa. 1981. *Impressões de viagem. CPC, vanguarda e desbunde: 1960/1970.* São Paulo: Brasiliense.
Castro, Hebe Maria Mattos de. 2006. "Remanescentes das comunidades dos quilombos: memória do cativeiro e políticas de reparação no Brasil." *Revista USP* (68): 104–11. http://www.historia.uff.br/culturaspoliticas/files/hebe1.pdf (accessed April 11, 2011).
Chagas, Mario de Souza. 2006. *Há uma gota de sangue em cada museu: a ótima museológica de Mário de Andrade.* Chapecó, Santa Catarina: Argos.
Chuva, Márcia Regina Romeiro. 2009. *Os arquitetos da memória: sociogênesis das práticas de conservação do patrimônio cultural no Brasil (Anos 1930–1940).* Rio de Janeiro: UFRJ.

136   *Myrian Sepúlveda dos Santos*

Dubois, Vincent. 1999. *La politique culturelle: Genèse d'une catégorie d'intervention publique.* Paris: Éditions Belin.
Dunn, Christopher. 2001. *Brutality Garden: Tropicália and the Emergence of a Brazilian Counterculture.* Chapel Hill: University of North Carolina Press.
Escobar, A. and Sonia Alvarez. 1992. *The Making of Social Movements in Latin America: Identity, Strategy, Democracy.* Boulder, CO: Westview.
Fraser, Nancy. 1998. "From Redistribution to Recognition? Dilemmas of Justice in a 'Post-Socialist' Age." In *Theorizing Multiculturalism,* edited by Cynthia Willett, 19–49. Oxford: Blackwell.
French, Jan Hoffman. 2009. *Legalazing Identities: Becoming Black or Indian in Brazil's Northeast.* Chapel Hill: University of North Carolina Press.
Freyre, Gilberto. 1933. *Casa Grande & Senzala: formação da família brasileira sob o regime da economia patriarcal.* Rio de Janeiro: Schmidt.
García Canclini, Nestor. 1990. *Culturas híbridas: Estrategias para entrar y salir de La modernidad.* Grijalbo: Mexico DF.
———. 1995. *Consumidores y ciudadanos: Conflictos multiculturales de la globalización.* Grijalbo: Mexico DF.
Gomes, Angela de Castro. 1994. *A Invenção do Trabalhismo.* Rio de Janeiro: Relume-Dumará.
Ianni, Octávio. 1968. *O colapso do populismo no Brasil.* Rio de Janeiro: Civilização Brasileira.
Instituto Brasileiro de Geografia e Estatísticas (IBGE). n.d. Censo 2010. http://www.ibge.gov.br/home/estatistica/populacao/caracteristicas_raciais/default_raciais.shtm (accessed September 30, 2011).
———. 2007. Sala de Imprensa. Síntese de Indicadores Sociais 2007. http://www.ibge.gov.br/home/presidencia/noticias/noticia_visualiza.php?id_noticia=987 (accessed September 30, 2011).
———. 2008. PNAD (Pesquisa Nacional por Amostra de Domicílios). http://www.ibge.gov.br/home/estatistica/populacao/condicaodevida/indicadoresminimos/sinteseindicsociais2010/SIS_2010.pdf (accessed September 30, 2011).
Instituto de Pesquisa Econômica Aplicada (IPEA). n.d. *Renda, Desigualdade, Coeficiente de Gini: 1976–2009.* http://www.ipeadata.gov.br (accessed September 30, 2011).
Laclau, Ernesto. 1977. *Politics and Ideology.* London: Verso.
Miceli, Sérgio. 1984. "Teoria e prática da política cultural oficial no Brasil." In *Estado e Cultura no Brasil,* organized by Sérgio Miceli 97–111. São Paulo: DIFEL.
———. 2001. *Intelectuais à brasileira.* São Paulo: Companhia das Letras.
Ministério da Cultura. n.d. *Sistema de Apoio às Leis de Incentivo à Cultura (Salicnet).* http://sistemas.cultura.gov.br/salicnet/Salicnet/Salicnet.php (accessed September 30, 2011).
———. 2010. *Texto-base da Conferência Nacional de Cultural.* http://www.cultura.gov.br/cnpc/wp-content/uploads/2010/08/texto-base_iicnc.pdf (access September 30, 2011).
Nederveen Pieterse, Jan. 2007. *Ethnicities and Global Multiculture: Pants for an Octopus.* Lanham, MD: Rowman & Littlefield.
Ortiz, Renato. 1986. *Cultura brasileira & identidade nacional.* São Paulo: Brasiliense.
———. 1991. *A Moderna Tradição Brasileira.* São Paulo: Brasiliense.
Peixoto, Paulo. 2000. "O Património Mundial como Fundamento de uma Comunidade Humana e como Recurso das Indústrias Culturais Urbanas." *Oficina do CES,* # 155.
Perrone, Charles. 1989. *Masters of Contemporary Brazilian Song: MPB 1964–1985.* Austin: University of Texas Press.

Ridenti, Marcelo. 2000. *Em busca do povo brasileiro: artistas da revolução, do CPC à era da TV*. Rio de Janeiro: Editora Record.
Said, Edward W. 1978. *Orientalism*. New York: Pantheon Books.
———. 1985. "Orientalism Reconsidered." *Cultural Critique* 1: 89–107.
Schwarz, Roberto. 1978. "Cultura e Política, 1964–1969." In *O pai de família e outros estudos*, 61–92. Rio de Janeiro: Paz e Terra.
Sovik, Liv. 2001. "Globalizing Caetano Veloso." In *Brazilian Popular Music and Globalization*, organized by Christopher Dunn and Charles A. Perrone, 96–105. Gainesville: University Press of Florida.
Stepan, Nancy. 1991. *The Hour of Eugenics: Race, Gender, and Nation in Latin America*. Ithaca, NY: Cornell University Press.
Veloso, Caetano. 1997. *Verdade Tropical*. São Paulo: Companhia das Letras.
Weffort, Francisco C. 1978. *O Populismo na Política Brasileira*. Rio de Janeiro: Paz e Terra.
Williams, Daryle. 2001. *Culture Wars in Brazil: The First Vargas Regime, 1930–1945*. Durham, NC: Duke University Press.
Young, Iris M. 1998. "Unruly Categories: A Critique of Nancy Fraser's Dual Systems Theory.' In *Theorizing Multiculturalism*, edited by Cynthia Willett, 50–67. Oxford: Blackwell.

8 Community Policing of Rio's Favelas
State-Led Development or Market-Oriented Intervention?

Erica Mesker

Favela Santa Marta (also known as Doña Marta) was given international attention in 1996 when Michael Jackson began filming the video for his international hit "They Don't Really Care about Us" within the favela. At the time, the government of Rio de Janeiro was staunchly against the filming and tried to block Jackson's arrival, arguing that it would bring negative attention to the city and hinder its attempts at rehabilitating its image. However, a higher court overturned the block and allowed filming to proceed. At the time, Santa Marta was still under control of the heavily armed drug gang *Commando Vermelho*. The *New York Times* reported that Jackson and his team collaborated directly with the traffickers to ensure the security of the pop star, even going as far as hiring gang members to act as security. The video became an international sensation, bringing "visibility to poverty and social problems in countries like Brazil without resorting to traditional political discourse" (Nagib 2003: 123).

Over a decade after Jackson's visit, on December 19, 2008, 123 UPP officers led by Captain Priscilla de Oliveira entered favela Santa Marta after a one-month occupation by the *Batalhão de Operações Policiais Especiais* (Police Special Operations Battalion), more commonly known by its acronym, BOPE. Developed with an Olympic bid in mind, the *Unidades de Polícia Pacificadora* (Pacifying Police Units), also known by their acronym UPP, became the crux of the security plan proposed to the 2016 Olympic Committee Evaluation Commission. According to Carlos Arthur Nuzman, head of Brazil's Olympic Committee, the Rio 2016 security proposals would "ensure the execution of the Olympic games with precision and technical excellence, and provide a legacy for the transformation of the city and country" (Rio2016 2009). Furthermore, *Billboard* magazine, in its July 11, 2009, issue, attributed the success of the UPP to the attention Jackson brought to the favela with a quote from Claudia Silva, press liaison for Rio's office of tourism, saying that "this process to make Dona Marta better started with Michael Jackson. Now it's a safe favela. There are no drug dealers anymore, and there's a massive social project. But all the attention started with Michael Jackson." Indeed, the introduction of the UPP has already begun transforming occupied favelas since its first installation in Santa Marta over four years ago.

The UPP are part of the state government's response to the chronic violence associated with drug trafficking in Rio de Janeiro's favelas. The UPP have garnered much local and international media attention and are widely seen as a successful attempt at establishing state presence within strategic favela communities that had long since been abandoned and abused by the same government and police force that is now charged with its security. Indeed, Rio de Janeiro's newest community policing model has helped curb violence in occupied favelas; however, new research presented here will demonstrate that although the UPP have the potential to act as a catalyst for social emancipation and equality, the deeper issues of poverty, social exclusion, and discrimination have been largely ignored. Moreover, when looking at the contradictory actions of the state it becomes clear that although the UPP have brought benefits to occupied favelas, the real beneficiary is market-oriented capitalist development.

THE FAVELAS OF RIO DE JANEIRO

Rio de Janeiro is Brazil's second-largest city with over 11.5 million people living in the metropolitan region of Rio de Janeiro. Rio is the most internationally recognized Brazilian city and the most visited tourist destination in South America (Mesker 2012). The skyline is instantly recognizable with its towering peaks and Christ the Redeemer, bathed in white, watching over the city below with outstretched, welcoming arms atop Corcovado Mountain. Also working their way up the precarious mountain slopes are Rio's favelas, which have become as well known as the posh beaches of Copacabana, Ipanema, and Leblon below. Of the 11.5 million people living in Rio, nearly 20 percent live in favelas. Because many favelas are illegal and outside of the formal city planning structure, they often lack basic municipal services. Despite the lack of services, Rio's favela population is growing. In fact, the growth rate of the favelas is over three times higher than the growth rate of the city itself (see Table 8.1).

Table 8.1   Rio de Janeiro Favela and City Population and Growth Rates, 1950–2010

Year	Favela Pop. (a)	City Pop. (b)	a/b (%)	Growth Rate: Favelas (%)	Growth Rate: Rio Pop. (%)
1950	169,305	2,337,451	7.24	-	-
1960	337,412	3,307,163	10.20	99.3	41.5
1970	563,970	4,251,918	13.26	67.1	28.6
1980	628,170	5,093,232	12.33	11.4	19.8
1990	882,483	5,480,778	16.10	40.5	7.6
2000	1,092,958	5,857,879	18.66	23.9	6.9
2010	1,393,314	6,320,446	22.04	27.5	7.9

Source: Perlman (2003) and IBGE (2010).

Rio's favelas became internationally recognized due to their extremely high levels of violence. In fact, Rio's favelas have a homicide-by-gun rate of 240 per 100,000—comparable to a country in a full state of war (Lucas 2008 in McRoskey 2010: 94). For example, between December 1987 and November 2001, 467 minors were killed in the conflict between Israel and Palestine due to firearm-related deaths, whereas during the same period, 3,937 minors were killed in the Rio de Janeiro municipality alone (Dowdney 2003: 172 in Brähler 2010). Moreover, because the vast majority of these deaths occur among the poor, black, male youth—who generally reside in favelas—former national secretary of public security Luíz Eduardo Soares has referred to the phenomenon as "genocide" (Soares, Bill, and Athayde 2005: 247 in Brähler 2010).

A 2008 study showed that 96.3 percent of the territory encompassing Rio's favelas was outside of state control (NUPEVI in Brito 2009; Brähler 2010). Most of the violence is attributed to the various rival drug gangs—and more recently, the militias—who fight between each other over territorial claims and against the Militarized Police (*Polícia Militar* [PM]). The following examples demonstrate the daily gang and police-related violence (Brähler 2010):

- Gang invasion of rival territory and other intergang conflicts
- Armed confrontations between drug gangs and the police
- Stray bullets, which frequently kill uninvolved and innocent bystanders
- Extrajudicial executions and torture
- Related arms crimes, for example the trafficking of weapons

Usually the government combats drug gangs by sending in the elite and oppressive Police Special Operations Battalion (*Batalhão de Operações Policiais Especiais*, more commonly known as BOPE) featured in the award-winning Brazilian film *Elite Squad*, and arguably the best trained urban fighting force in the world. BOPE enters the favela, finger on the trigger, searching for drugs, weapons, and the people who traffic them. The usual result is an intense battle between the police and the ruling gang, leading to unintended (and sometimes intended) civilian deaths. As summarized by Human Rights Watch (2011):

Police abuse, including extrajudicial execution, is a chronic problem. According to official data, police were responsible for 505 killings in the state of Rio de Janeiro alone in the first six months of 2010. This amounts to roughly three police killings per day, or at least one police killing for every six 'regular' intentional homicides. (215)

Stray bullets are common, as are 'deaths due to resistance'—a loophole that allows the PM to commit random acts of violence under the guise of resistance. The introduction of the UPP, however, represents a much needed move into the less repressive form of policing—community policing.

POLICING, BRAZIL STYLE

History and Organization

Efforts to police Rio's favelas have historically involved the traditional militarized model of policing. Historically, the police of Brazil were given the task of keeping slaves and recently freed men under control, usually by oppressive means (Bretas 1997; Holloway 1997). After Brazil was declared a republic in 1889, a year after slavery was abolished, the police maintained their violent and repressive tactics, especially against political enemies and the poor. This persisted through two periods of dictatorship (1930–34 and 1937–45) and a period of military rule (1964–85) (Da Silva and Cano 2007). After the end of the military rule, the 1988 Federal Constitution, as described in Article 144, divided the police of Brazil into three separate entities: the Federal, State, and Municipal (Figure 8.1). As summarized by

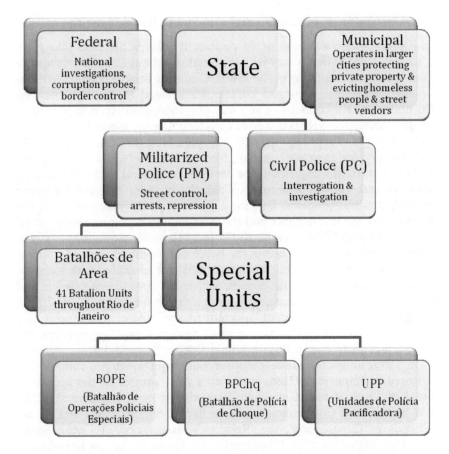

Figure 8.1   Breakdown of Brazil's policing hierarchy (source: Amar 2012).

142   *Erica Mesker*

Paul Amar (2012), "The Federal Police conduct national investigations, probe corruption, and control borders. The state/provincial police forces include the Militarized Police (*Polícia Militar*, PM), charged with street control arrest and repression, and the Civil/Judicial Police (*Polícia Civil* [PC]), charged with interrogation and investigation. The newer Municipal Guards, who operate in the larger cities on a questionable legal basis, protect private property and evict homeless people and street vendors."

The UPP is housed under the umbrella of the Militarized Police of the State of Rio de Janeiro (*Polícia Militar do Estado do Rio de Janeiro* [PMERJ]) along with BOPE and the Riot Control Battalion (*Batalhão de Polícia de Choque* [BPChq]), the repressive counterparts of the UPP, both of which play a role in the initial stages of UPP implementation, specifically invasion and stabilization. Although not a part of the Brazilian military, the PM is referred to as militarized because of its militaristic principles of hierarchy, weaponry, uniform, discipline, and ceremony.

The Development of Community Policing in Rio de Janeiro

In Rio de Janeiro a change in the politics of policing began to take place after the return to democracy with the election of Lionel Brizola as the state governor at the beginning of the 1980s. Governor Brizola, a human-rights-oriented politician, advocated communication and cooperation between the police and the community as a way of resolving problems in the public security sector, as well as a way of counteracting the consequences of the military regime's repressive tactics (Mesquita Neto 1999 in Saima 2007). In his second term as state governor, from 1991 to 1994, Brizola appointed Colonel Nazareth Cerqueira as the commander general of the PM and the state secretary of public security. Colonel Cerqueira reflected Governor Brizoa's progressive ideals, favoring a public security policy based on the demilitarization of the police and the introduction of community-oriented policing (Hinton 2006 in Saima 2007). In 1991 a pilot community-oriented policing project was launched in the beachside neighborhoods of Copacabana and Leme. The project was short-lived as it coincided with a drastic increase in crime in Rio de Janeiro, which led government officials and community members to question the efficacy of the program (Saima 2007). Moreover, with the election of Governor Marcello Alencar, a conservative who advocated a return to conventional, repressive policing practices, the project was abandoned (Oliveira Muniz 1999 in Saima 2007).

In 1999, under then state governor Anthony Garotinho, community-oriented policing resurfaced, again as a pilot project in the neighborhood of Copacabana. Governor Garotinho brought together the 19th PM battalion of Copacabana and community organizations such as the 'Friends of Copacabana', the 'Resident's Association of Copacabana', and the 'Association of Businessmen of Copacabana' to introduce a community-oriented policing strategy that would increase communication and cooperation between the police and community, as well as make the police more involved and available to residents, workers, and national and international tourists alike.

With the introduction of the community policing pilot project in Copacabana, "police officers who had once been distant figures symbolizing repression were visible on nearly every street corner, on foot, on bicycles, in community-oriented policing posts (booths manned by community-oriented policing officers specifically created in order to make the police accessible to the population at all times) in beach buggies, and sitting in cars" (Saima 2007: 134). The project received all-around support, from the police, the community, politicians, and the media, for securing this tourist hot spot. Although Copacabana's community policing project was—and still is—seen as a success, the favelas haphazardly meandering up Copacabana's surrounding hillsides remained in the blind spot of the state. However, in September 2000 the first ever community policing pilot program to be introduced within any Rio's favelas was implemented in the Pavão-Pavãozinho-Cantagalo favela between the neighborhoods of Copacabana and Ipanema to rid the area of the drug gang *Commando Vermelho* and bring it under state control (Saima 2007).

The aforementioned program, tactlessly named the Special Areas Policing Group (*Grupo de Policiamento em Áreas Especiais* [GPAE]), was created by former undersecretary of research and citizenship within the state secretariat for public security, Luiz Eduardo Soares, and coordinated by PM Major Carballo in cooperation with nongovernmental organizations and the church (Saima 2007). GPAE was initially seen as a success and was eventually expanded to include the favela communities of Vila Cruzeiro, Morro do Cavalão, and Morros da Formiga, Chácara do Ceu, and Casa Branca (Da Silva and Cano 2007; Saima 2007). Despite the initial success and expansion of GPAE, the project eventually came to a disastrous ending and established a precedent for police brutality and impunity. After the abandonment of GPAE, occupied favelas returned to the control of drug gangs, and the state subsequently returned to the repressive tactics of enter-shoot-retreat. Police brutality against favela residents became not only the norm, but the expectation, which typically consists of "targeting the poor and marginalized, entering favelas heavily armed and in force, using brutal and repressive tactics against its residents, and engaging in shoot-outs with suspected drug traffickers that often leave many civilians dead or injured" (Saima 2007: 140).

After decades of neglect and repression, and one failed attempt at community-oriented policing within a handful of Rio's favelas, the state is again experimenting with community policing. With the Olympics and the FIFA World Cup on the horizon, the government, at the local, state, and federal level, is heavily invested in the success of the UPP program.

THE PACIFYING POLICE UNITS OF RIO DE JANEIRO

Created under the administration of Rio de Janeiro state governor Sergio Cabral and State Secretary for Public Safety Jose Beltrame, the UPP were developed as a branch within the Militarized Police of the State of Rio de Janeiro (PMERJ) via decree number 41.650 on January 21, 2009. Unlike the infamously repressive tactics of the PM, the UPP represent a shift towards more inclusive tactics. The primary goal of the UPP is to establish state control over the use of force by occupying territories (favelas) that for decades have been under the influence of gangs[1] and, more recently, militia.[2] The establishment of a state presence allows for the delivery of basic services such as electricity, trash disposal, and health and educational services. Additionally, because most favela communities have experienced years of neglect, indifference, corruption, and abuse at the hands of police, the UPP focus on building trust between the population and the police.

The UPP Implementation Process

Favelas that are chosen to be pacified have been under the control of drug gangs for years, sometimes decades. The UPP are not equipped to expel drug traffickers, which is why four phases have been set in place to prepare the community for the UPP occupation (see Table 8.2).

Table 8.2   Phases of UPP Implementation

Phase	1	2	3	4
Pacification Process	Surveillance & Invasion	Stabilization & Occupation	Definitive Occupation	Postoccupation
Actors	BOPE, BPChq, occasionally Brazilian Armed Forces	BOPE, BPChq, occasionally Brazilian Armed Forces	UPP	UPP, Social UPP, Instituto de Segurança Pública (ISP)
Characteristics	• Announced war • Invasion with the use of tactical force, heavy weaponry & armored vehicles	• Thorough search and seizure of drugs, weapons, stolen goods and money • Maintenance of control	• Focus on community building • Increase of officer visibility & approachability • Introduction of social and community services	• Deepening of community relations • Extension of social & community services • Development of quantifiable benchmarks & studies to measure outcomes
Level of Repression	Very high	High	Low	Very low
Level of Prevention	Very low	Low	High	Very high

Source: Brähler (2010).

An important factor to take note of is the fact that the PM announces days ahead of time which favela will be pacified. This gives residents time to prepare for the entry of BOPE, and also gives traffickers time to flee. Like the majority of BOPE invasions, this first step of pacification can be violent, especially when traffickers decide to put up at fight.

After BOPE invades, a sweep of the favela is conducted to locate *traficantes*, drugs, weapons, stolen goods, and drug money. Residents of occupied favelas have reported cases of police theft from the homes of innocent civilians, as well as cases of police brutality and harassment. Despite complaints concerning theft, brutality, and harassment on the part of BOPE officers, residents believe that the invasion is a necessary part of the introduction of the UPP. A 2009 survey conducted by *Fundação Getuilo Vargas* (FGV) in the pacified favelas of Santa Marta and Cidade de Deus found that "74% of the residents in Santa Marta and 82% of the residents of Cidade de Deus were convinced that the implementation of the UPP is only possible with the foregoing police occupation by BOPE" (Brähler 2010).

Once the occupied favela is stabilized by BOPE, and in some cases, the Brazilian military,[3] the UPP enter to begin their patrols. In addition to patrolling the favela, officers also conduct interviews to find out what social services residents want most, and work to build some level of trust between the police and the community. Trust building can be achieved through increasing officer visibility and maintaining permanent patrols, as well as introducing community-building initiatives such as youth athletic and educational programs.

STRENGTHS AND SHORTCOMINGS OF THE UPP

Reduction in the Visibility of Drugs and Violence

One of the greatest successes of the implementation of the UPP is the decrease of drug trafficking and the associated violence. Victims of 'death due to resistance' has fallen sharply in UPP occupied favelas and the surrounding areas (ISP 2011). A 2010 study by the *Instituto Brasileiro de Pesquisa Social* (IBPS), in which researchers interviewed 600 residents from varying UPP occupied favelas, showed that the majority of residents had the impression that drug sales had ended (62 percent) or decreased (30 percent). More than 80 percent of the residents also stated that executions and shootouts had ended (IBPS 2010). Furthermore, 1,200 residents were interviewed throughout favelas Santa Marta and Cidade de Deus in the 2009 FGV study, in which 58 percent of the interviewees in Santa Marta and 64 percent in Cidade de Deus indicated that personal security had improved or improved a lot since the installation of the UPP.

These successes have led to the near disappearance of open-air drug markets and corners patrolled by assault-rifle-wielding traffickers. This will have an important, yet nonquantifiable impact on the social fabric of a pacified favela. Instead of children growing up among the persistent and somewhat idealized drug culture, looking up to those in charge and sometimes striving to be like them, children now see legitimate police officers on community patrols, usually by foot, gun holstered. That is not to say that drug dealing and gun toting has stopped all together—it certainly continues in pacified favelas—but it has become more clandestine, such as it is in many poor, middle-class, and upper-class neighborhoods throughout the world.

Spreading—Not Solving—the Problem

Although violence reduction in pacified favelas is a major success, a shortcoming of the program is that violence has been transferred to the periphery of the city as well as to the surrounding metropolitan regions. A persistent complaint is that, because of announced invasions, traffickers are able to flee from favelas scheduled for pacification. Many traffickers fled to Complexo do Alemão, but with its invasion and subsequent military occupation, along with the 2011 BOPE invasion and 2012 UPP occupation of favela Rocinha, many traffickers have fled to the larger Rio metropolitan areas such as Niterói, São Gonçalo, and Marica. Whereas the numbers of violent deaths in the state of Rio de Janeiro fell 23.5 percent, Niteroi and Marica had a 46 percent rise in these incidences from February 2010 to February 2011. The collective number of robberies in Niteroi and Marica jumped 80 percent in February 2011 compared to the same period last year. In these two municipalities, the cases of vehicle theft increased 31 percent and assaults on businesses rose 36 percent (Pereira 2011). Moreover, favelas outside of the city of Rio de Janeiro, which were until recently free of territorial drug gangs, are now under the strict rule of traffickers.

Although the goal of the UPP has never been explicitly aimed at ending drug trafficking or crime, nor has it been promised as a solution for all favelas and all socioeconomic problems in Rio, the UPP have certainly had a positive impact on occupied favelas. Unfortunately, whereas pacification has alleviated violence in UPP occupied favelas and the surrounding areas, this violence has not disappeared, but rather, has migrated away from the Olympic Zones and the tourist corridor to other less visible areas of Rio and the larger metropolitan region. Similar to community policing programs of the past, this has led to speculation that the UPP program may wind down after the conclusion of the 2016 Olympic Games, leading to a return to the rule of the traffickers. Indeed, even the UPP officers are skeptical of their staying power; in a recent survey, 70 percent of the officers interviewed felt that the program was created only to secure the city for the World Cup and Olympics (Soares et al. 2010: 16). It is no wonder why favela residents, as well as the officers themselves, remain cautious about the staying power of the UPP.

Staying Power of the UPP: GPAE and the Fear of Repeating the Past

One of the most persistent concerns is the staying power of the UPP. There are several reasons why after three years of relatively successful implementation people continue to doubt the staying power of the UPP. Fresh memories still remain of community policing initiatives of the past gone terribly wrong. Although many media outlets have touted the UPP program as revolutionary, it is not the first time community policing has been implemented in Rio's favelas.

In May 2000, in an example of Rio's traditional style of policing, PM officers burst into the Pavão Pavãozinho favela and executed five young men who the police claimed were drug traffickers. Infuriated residents descended into the wealthy beachside neighborhoods, breaking windows and setting fire to cars and busses. This was the catalyst that set a community policing experiment into motion. Former public-security adviser, Luiz Eduardo Soares, with Major Antonio Carlos Carballo, had been trying for years to convince Rio de Janeiro state governor, Anthony Garotinho, to implement a program for community policing. Despite political pressure from Rio's conservative city mayor, Cesar Maia, Governor Garotinho allowed a community policing plan to go forward. In September 2000 the Special Areas Policing Group (*Grupo de Policiamento em Áreas Especiais* [GPAE]) was launched in Pavão-Pavãozinho-Cantagalo favela, between the neighborhoods of Copacabana and Ipanema, after an announced invasion and subsequent stabilization by BOPE. Reminiscent of the media coverage surrounding the UPP, in a September 5, 2001, article in *The Economist*, Major Antonio Carlos Carballo pointed out bullet holes sprayed across a wall and proclaimed, "That's how things used to be here."

The first few years of the GPAE program were considered a success. In occupied favelas, shootings and fatalities were near zero and stray bullets became unheard of. However, due to changes in the command structure, the slow implementation of promised social projects, and corruption in the ranks, there was an eventual breakdown of community-police relations, culminating in several bloody massacres throughout 2004, which resulted in numerous deaths and multiple reports of torture committed by police (Da Silva and Cano 2007). The massacre was rapidly followed up by a quick trial, further establishing police impunity (Lyra et al. 2001: 37–42). Finally, after nearly seven years, the GPAE program, like so many other Brazilian initiatives, came to a halt with the changing government administration, while *Commando Vermilho* returned to previously occupied favelas with a vengeance against those who had cooperated with police.

The initial success story of GPAE is incredibly similar to that of the UPP; the reduction in gang-related violence and stray bullets, the positive media attention, and the excitement over a new, less repressive community policing program. Many of the weaknesses are also beginning to appear in the UPP, such as changes in the command structure and the slow roll out of social services. Also, despite UPP officers being paid an extra R$500/month on top of normal salary, 60 percent of UPP officers say that their income is insufficient (Soares et al. 2010: 13). This is an indicator that, as seen in a UPP corruption ring involving the former commander Cpt. Elton Gomes Costa, former deputy commander Lt. Rafael Medeiros, and thirty officers in the UPPs of Morros do Coroa, Fallet, and Fogueteiro in Santa Teresa, the more lucrative business associated with corruption, including accepting bribes and aiding traffickers, is still a serious problem that has not been overcome. According to Human Rights Watch (2011), "reform efforts have fallen short because state criminal justice systems rely almost entirely on police investigators to resolve these cases, leaving police largely to police themselves . . . The state has not yet taken adequate steps to ensure that police who commit abuses are held accountable" (115–16). Additionally, according to RioRadar, "research by Candido Mendes University discovered that only 40.6% of UPP officers are satisfied [working in the UPP] and 70% would prefer to be transferred to a normal battalion. Recent news stories have also highlighted that officers are complaining of the lack of proper training, uniforms, bullet-proof vests, weapons, and facilities. Most troubling is that the promised entrance of accompanying social services has been slow and substandard—many top officials acknowledge this, to varying degrees" (RioRadar 2011). Even Jose Beltrame agreed, noting that the purpose of the UPP is to "facilitate the arrival of dignity for the citizen," adding that "the project's success depends on massive investment, and these are not being made quickly enough" (Bottari and Gonçalves 2011).

Failure to Target Deeper Prejudices

Another reason for lingering questions about the staying power of the UPP is the contradictory approach the state is taking when dealing with Rio's favelas as a whole. Whereas deaths due to resistance have dropped to zero or near zero in pacified favelas, deaths due to resistance remain in the hundreds throughout the rest of Rio, even increasing in some areas (ISP 2011). According to Liz Leeds, a police reform specialist, in an interview with Rio Real (Michaels 2011):

> If community-oriented policing doesn't become an integrated policy of the police across the board, not just in favelas, it will ultimately not take hold. There's still a lot of resistance. Until the training really changes to get away from a very militaristic training and viewpoint of what the police are really about, disrespect toward communities, and thus a lack of cooperation from those communities, will remain a problem.

There is, unfortunately, a gap being created between pacified and non-pacified favelas, with residents of pacified favelas being treated with more respect and dignity by officers trained in community policing strategies and human rights. Not only does the pacified favela get its own special police force, but also its officers are trained on how to police humanely. Why is it that residents of nonpacified favelas must live in a daily struggle between the crossfire of traffickers and police? To someone in the middle class reading *O Globo*, human rights training seems like a positive step forward, while international news outlets pick up on the story and tout the progressive steps Rio has taken in training their officers in the field of human rights. In fact, according to a survey of UPP officers almost half of respondents think the media portrays the UPP more positively than they actually are (Soares et al. 2010: 14).

Human rights training is certainly a step in the right direction, but why were courses in human rights even necessary? The fact that officers need extra training shows how deep the prejudices and inequalities run throughout Rio's police force and the population at large. Additionally, if the implementation of human rights training has been so successful for the UPP, why hasn't the military police implemented human rights training throughout every department? Apparently, citizens living in UPP-occupied favelas are given a veil of equality only when inside their community. Outside the occupied favela, though, prejudice and inequality continue to be produced and reproduced daily.

The Contradictory Case of Vila Autódromo

Although the Rio government is focusing on the positive benefits of the UPP and favela pacification and integration in general, it is taking a contradictory approach when dealing with favelas that stand in the way of Olympic development goals. Whereas the Brazilian media covers the successful implementation of UPPs, it has generally ignored the forced eviction of residents and the destruction of entire favelas—something that has not occurred since the military dictatorship. How can the government's steps to integrate favelas into the formal city be taken seriously when at the same time they are razing entire communities?

For example, the favela *Vila Autódromo* is slated for removal by 2013, forcibly evicting 939 individuals from their homes within the favela. Interestingly, according to *O Globo*, the government has been trying to evict residents since 1993, but only recently received permission when it included the area in its Olympic plans (Magalhães 2011).

The Brazilian government is contracting out the labor of building up the Olympic Park, the site of which will be at *Vila Autódromo*, to whichever company that will provide the lowest public expenditure and propose the best technical solution for the implementation of the park. As part of the deal, the contractor will also be in charge of maintaining the site for fifteen years, after which the property will be transferred to private investors. A training center is supposed to remain on site after the fifteen-year period, but in all 75 percent of the land will be marked for private development. As part of the "My House, My Life" program, current residents either will be reimbursed the value of their home or will be able to move into new government constructed homes in *Parque Caricoa*, all while multimillion-dollar condos pop up where their homes once stood.

It is contradictory actions such as these that leave people wondering if the UPP is a long-term project, or just a security investment for the mega-sporting events coming to Rio. It also begs the question of who is actually benefiting from the implementation of the UPP, which, although part of state-level development, may also represent "new forms of neoliberal authority in relation to institutions of law enforcement as they participate in the roll-back of the state through privatization, as well as the roll out of new forms of interventionism" (Amar 2012: 6).

Avalanche of Services: State-Driven Development or Market-Oriented Intervention?

Favelas are areas that have statistics that read like a perpetual war zone and finally, with the installation of the UPPs, some favelas are enjoying a long-awaited respite from the constant shootings and abuse, which, to be clear, is not a benefit—it is a basic human right that has long been ignored by the state. Security and peace within occupied favelas does allow the state to provide much needed services such as trash collection, education, and health services. More than half of the interviewees in a study confirmed that access to state and private services within their favelas had improved or improved a lot. For example, an IBPS (2010) study revealed that 7 percent more residents had electricity than before the pacification and there was a 6 percent increase in access sewerage and garbage collection, as well as a 9 percent increase in access to paved streets. Access to commercial services also increased, with a 12 percent increase in access to (legal) TV, and 17 percent to deliveries such as pharmaceuticals and food delivery.

It is evident that the UPP occupation has opened up a largely untapped market. If all favelas were pacified, the Rio municipal government would gain R$90 million in new property and service taxes. The president of the electric company Light estimated that the Rio economy could grow by R$38 billion through increased commerce and new jobs, and Light itself stands to gain US$200 million that is lost per year due to pirated electricity.[4] For example, in Santa Marta residents were given energy-saving refrigerators only if they signed up for paid electricity. In Complexo do Alemão the banks of Bradesco, Caixa Econômica Federal, and Banco do Brasil have plans to open up branches within the favela. There is even a cable package being marketed specifically to occupied favelas by cable operator Sky called 'Sky UPP', which costs R$44.90 for eighty-nine channels and has been a huge success (Carneiro 2010). Despite these additional services, a rapidly growing concern among residents, government and police officials, and nonprofit workers is the slow to nonexistent rollout of social services and employment opportunities to fill the vacuum left by the fleeing drug gangs. Does increased access to consumer services alone lead to social emancipation and equality, or will it further the divide between the asphalt[5] and the favela?

Legitimate services, as well as a 400 percent increase in housing prices (*O Globo* 2010) have only intensified the social inequalities within occupied favelas. The Brazilian national monthly minimum wage was recently increased to $R545 (Simoes and Bristow 2011), not nearly enough to cover increased rent and the cost of services such as gas, water, and electricity, which before pacification were often obtained illegally, but were free. Just as any society has class rankings, so does a favela, and an increased cost of living only serves to price out those who can no longer afford to maintain their lifestyle. Entrepreneurs are certainly doing well in the newly opened favela economy; however, lower-income families are all but banished from pacified favelas into more distant favelas, which means changing schools for kids and longer commutes for those working in the center of Rio. Whereas low-income families are pushed further to the periphery along with the violent drug gangs, the pacified favelas, with their new services and safety, have become havens for university students, artists, international travelers, and young professionals who welcome the opportunity to live in the city center at a lower cost while enjoying the best views Rio has to offer.

CONCLUSION

The UPP program has received a warm welcome in Rio and from the international community as a whole. To the extent that UPP-pacified favelas have reduced violence the program is no doubt a success. However, questions remain surrounding the long-term sustainability and staying power of the UPP as well as the underlying motives. Lack of personnel, equipment, and adequate funding, as well as a less than enthusiastic acceptance of community-oriented policing throughout the police subculture of the Militarized Police, in addition to instances of UPP officer corruption, leads one to question if the UPP will eventually succumb to failure, similar to community policing programs of the past.

The contrasting approaches of the state and Rio's Militarized Police in relation to pacified versus nonpacified favelas also leads one to question the staying power and motives of pacification. Thus far, only favelas in the tourist corridor and Olympic Zones have been pacified—a harsh, yet logical reality—and although police in these areas have received specialized human rights training there are still hundreds of favelas throughout Rio de Janeiro, as well as the surrounding metropolitan area, that are still victim to police brutality and repression on a daily basis. Human rights training and community-oriented policing practices should be implemented department-wide, not just among UPP officers.

The role of the private sector is also an interesting factor in the success of the UPP. Pacification has provided safe and open access to select favelas, opening up pacified favelas to the entrepreneurial spirit and free-market capitalism. Yet access to improved social services, such as better education and health care, is still slow in coming. Shootouts and homicides have dropped drastically; however, the UPP experiment in community policing needs to be successful beyond just its security aspect—it needs to be the catalyst for social emancipation and equality. To achieve this, the deeper issues of poverty, inequality, discrimination, and inadequate resources and opportunities, all of which lead people into the violent drug trade, need to be overcome. Without addressing these deeply rooted socioeconomic inequalities and questioning why it is that people enter into the drug trade, violence and marginality will persist. The UPP may provide security for select favelas, but without tackling the root of the problem, issues of violence, poverty, and inequality will not be solved, but will be transferred, pushed out of view, into the periphery. And as the UPP push drug gangs, violence, and low-income populations further into the periphery, away from the tourist corridor and Olympic Zones, the middle class is afforded access to relatively safe and low-cost living with the most spectacular views of *a Cidade Maravilhosa* (the Marvelous City).

Although for some this may be a disturbing reality, it has become clear that the investment and buy-in of the private sector will likely be the main factor in predicting the long-term success of the UPP. If the private sector sees money to be made in this newly emerging economy, and if middle-class favela residents continue to demand services and safety, the UPP will survive for a longer period and hopefully expand its reach beyond the Olympic and tourist zones. However, if the Brazilian state, and society at large, continues in its failure to discuss and bring attention to the underlying issues that have created the social and economic gap between those of the formal city and those of the favelas, no policing strategy, militaristic or community oriented, will have a chance at long-term survival within the favelas of Rio de Janeiro.

NOTES

1. The largest of which are *Comando Vermillhio* (Red Command), *Amigos dos Amigos* (Friends of Friends), and *Treciero Commando* (Third Command).
2. Militias generally consist of retired or off-duty police officers, firefighters, security guards, and political figures, such as members of Congress.
3. In the case of Complexo do Alemão, the Brazilian Military maintained control for one and a half years, arguably creating a military state for residents, before turning control over to BOPE. UPP units are now slowly being introduced to specific communities within Complexo do Alemão.
4. See leaked US Cable 09RIODEJANEIRO329: Counter-Insurgency Doctrine Comes to Rio's Favelas.

5. The formal city is commonly known as *O Asfalto*, "The Asphalt," because unlike in the favelas the wealthy communities of the formal city have paved streets.

REFERENCES

Amar, Paul. 2012. "Policing Systems." In *Encyclopedia of Global Studies*, edited by Helmut Anheier and Mark Jurgensmeyer. Thousand Oaks, CA: Sage.
Bottari, Elenilce, and Liane Gonçalves. 2011. "Beltrame quer pressa em investimentos sociais pós-UPPs: 'Nada sobrevive só com segurança'" *O Globo*, May 28. http://oglobo.globo.com/rio/mat/2011/05/28/beltrame-sobre-upps-nada-sobrevive-so-com-seguranca-hora-de-investimentos-sociais-924557293.asp (accessed June 2, 2011).
Brähler, Verena. 2010. "'Avalanches of Snow' in Rio de Janeiro: An Analysis of the City's Different Approaches to Combat Drug Trafficking." Master's thesis, University of London, Institute for the Study of the Americas.
Bretas, Marcos. 1997. *Ordem na cidade: O exercito cotidiano da autoridade political no Rio de Janeiro: 1907–1930*. Rio de Janeiro: Rocco.
Brito, Diana. 2009. "Trafico ou milícia dominam quase todas as favelas no Rio, mostra pesquisa." Folha Online, November 10. http://www1.folha.uol.com.br/folha/cotidiano/ult95u650276.shtml (accessed May 12, 2011).
Carneiro, Julia Dias. 2010. "UPPs abrem caminho para economia formal em favelas do Rio." BBC Brasil, December 20. http://www.bbc.co.uk/portuguese/noticias/2010/12/101220_rio_upps_servicos_julia_rw.shtml (accessed May 16, 2011).
Da Silva, Graziella Moraes D., and Ignacio Cano. 2007. "Between Damage Reduction and Community Policing: The Case of Pavao-Pavaozinho-Cantagalo in Rio de Janeiro's Favelas." In *Legitimacy and Criminal Justice*, edited by Tom R. Tyler, 186–214. New York: Russell Sage Foundation.
Dowdney, Luke. 2003. *Children of the Drug Trade: A Case Study in Organized Armed Violence in Rio de Janeiro*. Rio de Janeiro: Letras.
Fundação Getuilo Vargas. 2009. *Avaliação do Impacto do Policiamento Comunitário na Cidade de Deus e no Dona Marta*. Rio de Janeiro: FGV.
Holloway, Thomas. 1997. *Policia no Rio de Janeiro: repressão e resisténcia numa cidade do século XIX*. Rio de Janeiro: Editora Fundação Getulio Vargas.
Hinton, Mercedes S. 2006. *The State on the Streets: Police and Politics in Argentina and Brazil*. Boulder, CA: Lynne Rienner Publishers.
Human Rights Watch. 2011. "Brazil." In *World Report 2011*, 215–21. New York: Human Rights Watch.
Instituto Brasileiro de Geografia e Estatística (IBGE). 2010. Cities 2010 Census Data: Rio de Janeiro. http://www.ibge.gov.br/cidadesat/topwindow.htm?1 (accessed May 12, 2011).
Isntituto Brasileiro de Pesquisa Social (IBPS). 2010. *Pesquisa sobre a percepção acera das unidades de polícia pacificadora*. Rio de Janeiro: IPBS.
Instituto de Segurança Publica (ISP). 2011. "Vítimas de Auto de Resistência, Série Histórica." http://urutau.proderj.rj.gov.br/isp_imagens/Uploads/AutosDeResistencia.pdf (accessed November 2, 2011).
Lucas, Peter. 2008. "Disarming Brazil: Lessons and Challenges." *NACLA* 41: 27–31. https://nacla.org/article/disarming-brazil-lessons-and-challenges (accessed May 10, 2011).
Lyra, Diogo Azevedo, Marcelo Freixo, Marie-Eve Sylvestre, and Renata Verônica Côrtes de Lira. 2004. "Rio Report: Police Violence and Public Insecurity." Centro de Justiça Global.

Magalhães, Luiz Ernesto. 2011. "Minuta de edital prevê remoção de favela Vila Autódromo até 2013 para obras do Parque Olímpico." *O Globo*, October 4. http://oglobo.globo.com/rio/minuta-de-edital-preve-remocao-de-favela-vila-autodromo-ate-2013-para-obras-do-parque-olimpico-2744502#ixzz1g9K-kC5Sf (accessed October 4, 2011).

McRoskey, Samantha R. 2010. "Security and the Olympic Games: Making Rio and Example." *Yale Journal of International Affairs* 5 (2): 91–105. http://yale-journal.org/2010/07/security-and-the-olympic-games-making-rio-an-example/ (accessed April 10, 2011).

Mesker, Erica. 2012. "Rio de Janeiro." In *Encyclopedia of Global Studies*, edited by Helmut Anheier and Mark Jurgensmeyer. Thousand Oaks, CA: Sage.

Michaels, Julia. 2011. "Let's Talk about the Relationship." Rio Real Blog, May 29. http://riorealblog.com/2011/05/29/lets-talk-about-the-relationship/ (accessed May 29, 2011).

Mesquita Neto, Paulo de. 1999. "Policiamento Comunitário: a experiência em São Paulo." *Revista Brasileira de Ciências Criminais, Instituto Brasileiro de Ciências Criminais, São Paulo* 7 (5): 281–292.

Nagib, Lúcia. 2003. *The New Brazilian Cinema*. London: I. B. Tauris

O Globo. 2010. "Imóveis em favelas com UPP sobem até 400%." May 29. http://oglobo.globo.com/rio/mat/2010/05/29/imoveis-em-favelas-com-upp-sobem-ate-400-916732643.asp (accessed December 17, 2010).

Oliveira Muniz, Jacqueline de. 1999 *'Ser Policial é, sobretudo, uma razão de ser': Cultura e Cotidianoda Polícia Militar do Estado do Rio de Janeiro*. PhD thesis at the Universidade Federal Fluminense.

Pereira, Fernanda. 2011. "Município de Niterói segue na contramão da queda da criminalidade." *Journal O Fluminense*, April 21. http://jornal.ofluminense.com.br/editorias/cidades/niteroi-segue-na-contramao-da-queda-da-criminalidade (accessed June 10, 2011).

Perlman, Janice E. 2003. "The Chronic Poor in Rio de Janeiro: What Has Changed in 30 Years?" *The Journal of Human Development*. http://www.worldbank.org (accessed June 4, 2010).

Rio2016. 2009. "Rio de Janeiro Government Expands Community Policing." http://www.rio2016.org/en/rio-2016-now/rio-de-janeiro-government-expands-community-policing (accessed October 15, 2011).

RioRadar. 2011. "Unidades de Polícia Pacificadora: The Honeymoon Is Over." September 28. http://rioradar.com/archives/737 (accessed November 30, 2011).

Saima, Husain. 2007. *In War, Those Who Die Are Not Innocent ('Na Guerra, Quem Morre Não É Innocente'): Human Rights Implementation, Policing, and Public Security Reform in Rio de Janeiro, Brazil*. Amsterdam: Rozenberg.

Simoes, Carla, and Matthew Bristow. 2011. "Rousseff Wins Senate Battle on Brazil Minimum Wage Rise." Bloomberg Online, February 23. http://www.bloomberg.com/news/2011-02-24/rousseff-wins-senate-battle-on-brazil-minimum-wage-rise.html (accessed June 10, 2011).

Soares, Barbara Musumeci, Julita Lemgruber, Leonarda Musumeci, and Silvia Ramos. 2010. "Unidades de Polícia Pacificadora: O Que Pensam os Policiais." Centro de Estudos de Segurança e Cidadania (CESeC).

Soares, Luíz Eduardo, M. V. Bill, and Celso Athayde. 2005. *Cabeça de porco*. Rio de Janeiro: Editora Objetiva.

9 Media Democratization in Brazil Revisited

Carolina Matos

INTRODUCTION

A core concern that has remained at the center of the public service broadcasting ethos has been the ways in which ideas, information, and debate can contribute to promote progress, assisting in national development and improving the health of a particular democracy. One of the key purposes of my book *Media and Politics in Latin America* (2012) was to precisely examine the state and the challenges posed to public service broadcasting (PSB) and the public media at the turn of the twenty-first century in Brazil and in Latin America in a comparative perspective to the 'crisis' of identity of public communication structures across Europe and in the UK due to various factors including increasing media commercialization, expansion of new technologies, and fragmentation of audiences.

This chapter provides a synthetic summary of the key intellectual debates and research findings of *Media and Politics in Latin America*. It contains a synthesis of the methodology employed and some of the main research findings obtained from the online survey conducted with Brazilian students from the Communication Department at UFRJ University in Rio de Janeiro.

Some of the questions asked included the potential and capacity of public communications to deepen the democratization project in Brazil. It also examined the ways in which the public media can offer better quality information and debate to larger sectors of the Brazilian audience independently of socioeconomic status, thus functioning as a unifying public sphere and assisting in the social inclusion of less privileged sectors of Brazilian society in national political debate. Here it is important to evaluate for instance the experiences of PSBs in European democracies, their historical and cultural relationship to democratization, and the lessons that can be learned from precisely this tradition of public service broadcasting in Europe and how they can be applied to the Latin American case (Matos 2012).

The four main lines of inquiry that I have pursued in this research have consisted firstly in comparing and contrasting the tradition of public service broadcasting in Europe to the situation, and the ways in which this has been directed to the public interest, to the authoritarian tradition of misuse of public communication structures for political purposes in Latin America and in Brazil. Thus another line of investigation closely connected to this is the further assessment of the exact nature of the relationship of public communications with the state, the public interest, and the public sphere. This leads to the third line of questioning, which is the examination of the debates on what constitutes 'quality' programming and information in both the private *and* public media, for the assessment of the tradition of PSB is tightly linked to issues of quality information and drama, in-depth political coverage, and accuracy, balance, and honesty in reporting. This is situated within the context of the 'crisis' of civic forms of communication and of political and 'serious' journalism in advanced democracies due to excessive commercialization, posing what many claim to be a tendency of lowering down quality standards (or 'dumbing down') as well as threatening media pluralism and the public sphere.

This chapter thus starts by examining briefly some of the challenges posed to the deepening of democracy and media democratization in Latin America, shifting the focus to look at the Brazilian case in comparative perspective to the UK and in greater depth. It further underlines the interweaving of economic and social inequality with the strengthening of political diversity and the struggle to advance media democratization in Brazil. The second half of the chapter provides an overview of the research methods employed in the research, highlighting some of the survey findings conducted with students on how they understand the public media and how it can contribute to the country's democratization project.

LATIN AMERICAN MEDIA IN COMPARATIVE PERSPECTIVE

The Latin American continent has changed significantly since the fall of dictatorship regimes. Democracy has slowly began to flourish in the continent amid the rise to power of center to center-left-wing governments in recent years, culminating in new approaches to foreign policy, new efforts of restructuring the state, expansion of internal and global markets, and the deepening of welfare and income distribution programs. Other innovations have included the adoption of initiatives aimed at empowering public communications to assist in the democratization process as a means of guaranteeing information rights to vast segments of the population independently of economic income and social status.

Political liberalization in Latin America has undoubtedly opened the avenue in the continent to revisit these debates on media democratization in a changed historical and political context that is a contrast to the dictatorship years of the 1980s, but that is nonetheless not entirely free from some of the dark clouds of the period, including the rise of accusations of the return to censorship practices and anxieties over press freedom. Nevertheless, governments across the continent not only are having to listen to the demands of civil society players, academics, journalists, and other members of the public in favor of a better and more accurate media, but are being pressured to formulate new media regulation policies capable of attending to the public interest. Democratic strategies are thus been envisioned as a means of reverting the region's current indicators of high media concentration and predominance of the market in the media sector (i.e., Moraes 2009).

Notably, the gradual democratization of Brazil and its social and political institutions in the last three decades has taken place not altogether disassociated from the authoritarian legacy that has marked the very formation of Brazilian society. As my last research has shown (Matos 2008), political liberalization and market expansionism in Brazil during the re-democratization period paved the way for the rise of journalistic professionalism in newsrooms. The improvements in the media and journalism during the 1990s, including wider commitments to equilibrium in political reporting during election campaigns, as well as the restructuring of key media industries, such as the newspapers *O Globo* and *Folha*, in order to better attend to multiple post-dictatorship publics (Matos 2008), are still far from being the main symbols of genuine media democratization. Moreover, the recognition alone that the media became more professional, including wider voices in the mediated sphere, is not a reason enough to state that the struggles for media democratization are a thing of the past. In many ways the fight has just began.

That a close relationship exists between media development, good governance, and the health of a democracy has been emphasized by various journalists, policy makers, and researchers (i.e. Schramm, 1964; Norris 2004). The 2010 UNESCO report, *Media Development Indicators: A Framework for Assessing Media Development*, underlined the close relationship that exists between the health, independence, and quality of the media with a country's development.[1] As Norris (2004: 1) has also argued, media systems can strengthen good governance and promote positive development, especially if there is a free and independent press capable of performing the watchdog role, holding powerful people to account, and acting as a civic forum of debate between society's competing interests.

However, a freer and more independent media and balanced press can operate only if they are not subject to either political or economic constraints (Hallin and Mancini 2004), and if public service media systems are also directed to serving the common good, and not misused for the personal interests of political and/or economic groups. It is no surprise that in Europe in liberal democracies the state's participation in the ownership or regulation of the broadcast media in liberal democracies has been largely based upon the need to guarantee standards of 'neutrality', minimizing political bias (Dunleavy 1987).

The UK for instance has managed to establish a sophisticated system of regulation and funding of PSB that has made it easier for broadcasters to be less obsessed with audience numbers and economic pressures, and thus more committed to serving the public. Set up under the 2003 communications bill, the UK's broadcasting regulator, *Ofcom*, has been an example of reference in media regulation in Europe, having defined a solid framework of regulation for British PSB.[2] *Ofcom* mainly requires UK public *and* commercial broadcasters (BBC, C4, ITV) to produce news with "impartiality and accuracy," with no editorial stances on political and controversial issues. This is seen as a means of guaranteeing an adequate degree of balance and fairness in the provision of news to the wider public (Ofcom, 2008a, 2008b).

As Dunleavy (1987) argues, public service broadcasting regulation in the UK has managed to act as a counterweight to the press, neutralizing or balancing the biases of the partisan British tabloids by offering more 'trustworthy' information. Its role in broadcasting is seen as one that is tightly connected to the public interest, as well as to the uses of the public media for educational and cultural purposes (Santos and Silveira 2007), further securing political coverage that is impartial between parties and which tends to privilege the collective good.

UK's PSBs have also been successful in fostering and mediating debate around the core issues of the day, providing a good balance of in-depth information and analysis with quality entertainment. British newspapers have grown under a tradition of editorial independence that has its roots in the struggles for press liberty and independence against monarchs and the state during the eighteenth and nineteenth centuries in Europe. Thus the British press has operated under a system of self-regulation largely represented through the Press Complaints Commission (PCC), a weak and largely inefficient body in contrast to *Ofcom*. Moreover, the notion of 'balance' in the press is understood differently than in broadcasting, and in the former it is largely perceived as being the end result of the competing views put forward by different newspapers in the marketplace. Nevertheless, calls for the statutory regulation of the press were raised emphatically in the aftermath of the *News of the World* phone-hacking scandal in June 2011, an issue that raises a whole new debate in the tradition of press self-regulation in the country.

However, the focus of this research is largely public service broadcasting as well as broadcasting regulation. Arguably, the literature on media democratization (Voltmer and Schmitt-Beck 2006; Curran and Park 2000; Sparks 2007) stresses how countries as different as South Africa, Chile, and China have encountered various problems in regards to the democratization of political communications. There were difficulties with implementing a more neutral, independent public service broadcasting (PSB) model similar to the UK's BBC in various new democracies for instance.

As Voltmer and Schmitt-Beck (2006) further underline, some countries in Eastern Europe have managed to implement PSBs with some degree of independence both from the state *and* from market competition. This is currently Brazil's main challenge. Thus at a moment when Brazil is worried about media democratization, countries like the UK are interested in preserving the tradition of PSBs in an increasingly uncertain future for European public service broadcasters and for organizations like the BBC (Scannell 1989; Raboy, 1996; Keane 2000; Curran and Park, 2000).

In the case of Latin America and Brazil, there are a series of global, national, regional, and local issues that need to be tackled that are closely interwoven with improvements in media systems and public communication structures. The 2004 report published by the United Nations Program for Development, *Democracy in Latin America: Towards a Democracy of Citizens*, talked to political leaders, business elites and entrepreneurs, academics, and forty-one presidents in order to assess the main obstacles to the consolidation of democracy in the continent. One important element cited were the tensions that existed between the institutional powers of the countries. The report nevertheless listed three main points, including internal limitations as a consequence of inadequate institutional controls and the multiplication of interest groups that functioned like lobbyists. The report also underlined external factors provoked by international markets, such as the threat posed by drug dealings as well as increasing media concentration. In answer to the question on who exercised more power in the region, the response was the financial economic sector (79.8 percent) and the commercial market media (64 percent).

After taking into consideration their historical differences, levels of economic and political development, power, and wealth, it is possible to underscore that democracies across the world in an age of globalization face similar democratic struggles in regards to inequality of income, economic deprivation, social exclusion of certain segments of society, and poverty as well as various other forms of taken-for-granted injustices. As Blaug and Schwarzmantel (1988: 1) note, several countries have not achieved the goal of becoming fully democratic states, encountering various difficulties in putting into practice the core values of democratic theory given the complexities of economic globalization and national politics.

As Held (1995: 3) states, democracy is associated with values of not only political equality, but liberty, common interest, self-development, and social mobility, or a means to legitimize the decisions of those voted into power. Democracy, continues Held (1995), needs to be deepened and extended both within *and* between countries, something essential if democracy is to claim its relevance in the centuries ahead. For democratic struggle, as Blaug and Schwarzmantel (1988) also assert, is above all about *expanding* the space for the inclusion of a wider citizen body, avoiding exclusions based either on property, gender, race, or ethnicity, which is a problem of both developed *and* developing societies alike. What differs is the degree and the extent of that inequality.

In this way the improvements in media systems in a particular country are closely interwoven with other betterments in the political sphere,

including a country's economic power and increase in quality levels of education and culture. This is precisely why the strengthening of the public
media platform is closely linked to the *development* and democratization
of a country's culture, as well as to the overall improvements in the *quality*
of its education and the number of people who have access to it. Thus it is
to some of the key achievements in media democratization in Brazil and in
other Latin American countries, as well as to the core obstacles that impede
further democratization of communications, that I turn to next.

CHALLENGES AND ACHIEVEMENTS

Brazil's authoritarian legacy has resulted among others in the marginalization of politics from the mainstream media. Television commercial broadcasting in the country to start with has been allowed to operate largely
unregulated (Lins da Silva 1990; Straubhaar, 2001). There has been a tendency to privilege entertainment and a consumerist aesthetic to the detri-
ment of more accurate and in-depth (political) debate. To start with, public
communication policies in Brazil date back to the period of the dictatorship
of the 1960s. This has occurred in spite of the fact that the progressive 1988
Brazilian Constitution emphasized in its key articles the need for a complex
media system composed of the state, public, and the private sectors, as well
as having introduced various articles concerning the need for regional and
independent production in the broadcasting field.

Debates on the necessity for further media democratization and the
updating of outdated laws, many of which were created before or during the dictatorship years, eventually culminated in the realization of the
much awaited *Confecom* (National Communication Conference) discussions in 2009. These were perceived as a direct result of the struggles and
pressures placed on governments and elites during the redemocratization
period by civil society representatives, journalists, and academics since
the 1988 Brazilian Constitution. Many Brazilian academics have underscored how the country has advanced less in media reform in contrast to
others in Latin America. The realization of the *Confecom* debates and the
implementation of *TV Brasil*, followed by the unification of various state
and educational channels, the granting of some funds to support regional
players, and the commitment assumed by the former administration in
favor of the creation of a new regulatory framework for the media, have
been some of the main achievements in media reform in the last eight
years of the Lula governments.[3]

Interviewed for this research, Cesar Bolano, professor at the Federal Ser-
gipe University and UnB, pointed out that a key demand of civil society is
simply to ratify the articles 220, 221, and 223 of the Brazilian Constitution. This to start with would begin to pave the way for media democratization. Notably, the first article prohibits the formation of monopolies

and favors press liberty; the second states that radio and television stations should prioritize educational, artistic, and cultural rationales; whereas the third declares that the private, public, and state systems should be explicitly contemplated as a means of guaranteeing the functioning of a proper complex media market that has all these sectors.

Bolano has also added that not much improvement has been detected in the restructuring of the public media platform in the country. In an interview given to the *National Forum of Communication Democratization* (FNDC), Bolano emphasized that the public media has still the same space as before: "What happened was a restructuring of the public television, but the public TV in Brazil still has the same space . . . in terms of audience share and effective production."[4]

The fact of the matter is that media organizations in Brazil still cultivate close ties to particular political parties, either directly or indirectly and regardless if they are private or public vehicles. According to former TV Cultura vice director of journalism, Gabriel Priolli, who was interviewed for my research in *Media and politics in Latin America*, the idea in favor of the public media was already subjected to politics since its very birth:

In 2005, when the *mensalao* scandals emerged, that was when they 'sold' the idea to Lula to have *TV Brasil*, of having a strong public network capable of competing with the private, as the government wanted a media which could be more favourable . . . The government wanted an instrument to defend itself, and it convinced itself that it was important. This is a contradiction with the real role that public TV should have. . . . There is actually a lot of idealism and hypocrisy in this whole discussion . . . People say that all you need is another option to *TV Globo* for people to change channels, but the reality is that they do *not*, they do not change to *TV Brasil*. I believe that this issue has a direct relation to education as well, for a better quality education produces audiences of better quality.

The public media sector in Brazil thus suffers from various historical deficiencies. It is composed mainly of the respected but funding-starved *TV Cultura* in SP and its counterpart *TVE* in Rio, as well as other regional outlets controlled by local politicians and sectors of the evangelical Church.[5] The community channels are broadcast on cable television (i.e., TV Senado, etc.), whereas the educational stations are in the hands of state governors. The current Brazilian TV market that is funded with public resources includes the television stations *TV Cultura*, which has an annual budget of R$160 million; *Radiobras*, with R$100 million; and *TVE*, which had R$35 million in 2004 and which has been incorporated into *TV Brasil*. There are also other resources that go to the television stations of the legislative federal, state, and municipal powers, plus *TV Justica* and university channels (Possebon 2007: 290), all of which have a low audience rating. The main media players in Brazil nonetheless—*Globo, Record, SBT, Bandeirantes,* and *Rede TV!*—have 82.5 percent of the national open television audience, of which 53 percent of the public are composed of people from the so-called class C (low middle class).[6]

The total funding for *EBC* includes money from the federal government as well as donations. According to the former minister of communications, Franklin Martins,[7] the new channel received a budget of R$350 million. The main programming is provided by Rio's educational television (*TVE*), with two programs from *Radiobras*. The morning slot is largely dedicated to children's shows as well as distant-learning programming. *TV Brasil's* programming also consists of hourly independent and regional programs, including the famous high-brow talk show *Roda Viva* and the journalism program *Jornal da Cultura* from *TV Cultura*, which is being retransmitted by *TV Brasil*.

After conducting a seminar with regulators and experts from across the world on the topic, in December 2010 however, one year after the *Confecom* debates, the Brazilian government announced its intention to implement new and updated media regulation policies, which were put on hold and given to the government of Dilma Rousseff (2011–14) to evaluate. The Ministry of Communications of the Dilma government hinted at the possibility of establishing two communication agencies.[8] *Anatel* would continue monitoring technical aspects, whereas the other agency would be created to ensure that the articles of the Constitution are respected. Little advanced in the first six months of the Rousseff administration, beyond the debates on abandoning controversial terms such as 'control of the media' in texts and documents. It clearly seems that debates on the formulation of new broadcasting regulation and media reform will occupy practically most of Dilma's mandate in office.

Without a doubt, the politicization of broadcasting, and the relationship established between media sectors with governments and the state, varies from country to country. It is dependent on historical and cultural factors; the degree of partisanship of the media; the size and power of the commercial press; and the extent to which journalists operates within a relatively strong regime of press freedom (e.g., Hallin and Mancini 2004). Notably, Brazil's reality can be considered more similar to the Argentine and Chilean cases in terms of the existence in the country of a stronger commercial press, media independence, and relative press freedom. In common with these other countries, journalism in Brazil has been engulfed in a history of censorship and struggle for stronger editorial independence, and the redemocratization years have seen the mainstream commercial press deal with the complexities of creating a more professional journalism culture to attend to the needs of the country's multiple publics and interests (Matos 2008).

Argentina for example is seen as being a contrast to the Brazilian case. In the latter country, the powerful lobby of *TV Globo* and of other market liberals is currently being considered a major impediment to further media democratization, amid other fears of limits to press freedom due to new regulation policies. Nonetheless, the media reforms that have been carried out in Argentina have been signaled out by academics and others in Brazil as having been in overall positive. In October 2009, the new audiovisual communication services law was sanctioned, substituting the legislation from the dictatorship period. The new law establishes some limits on media concentration, with each firm not being able to have more than ten radio and TV stations. It also authorizes the creation of the Federal Council of Communications, establishes quotas for local production, and poses limits on foreign participation in the firms of the sector in 30 percent. This has not occurred nonetheless without clashes between the biggest media group in the country, *Clarin*, and the Kirchner government.

Moreover, in Ecuador media reform debate has reached center stage since 2009. Discussions have emerged concerning the establishment of a Communication Council to regulate content, whereas in Venezuela, in spite of the creation of the international channel, *Telesur*, the Hugo Chavez government is being accused of power abuse. It has denied the renewal of the concession for the most popular and oldest channel in the country, *RCTV*, accused of supporting Chavez's coup in 2002.

Thus similar to the critical deliberations regarding the democratic potentials of the public sphere, which can be provided by new technologies, the public media sector can be seen as being capable in developing countries of being much more pronounced and more committed to the public interest than it currently is. It has the capacity of boosting political pluralism, while assisting also in the development of educational and cultural levels and granting this access to wider sectors of the population.

Another important point to emphasize in this debate is that education should not be disassociated from the politics of communications. It is evident that both are tied together and mutually interdependent. Thus the clear purpose of improving the quality of education in the country, of boosting cultural levels to wider sectors of the population, should be clearly connected to media democratization and media reform. Thus judging from the responses of the UFRJ survey and the interviews with experts, it is clear that sectors of the public in Brazil are interested in media improvements and quality programming, and are open to a more balanced combination of in-depth debate and information with entertainment, as we shall see next.

METHODS AND EMPIRICAL WORK

My current research has made use of a sophisticated triangulation methodology. This includes the application of an online survey to 149 communication students at UFRJ university in Rio de Janeiro, Brazil; the conduction of in-depth interviews with 12 policy makers, journalists, and politicians;

the discussion of programs from the public TV channel *TV Brasil*; as well as the critical assessment of the impact and uses of party political websites and blogs during the 2010 Brazilian presidential elections campaigns, which saw the election of the country's first woman president, Dilma Rousseff.

Notably, the difficulties in Brazil with implementing further media reform are rooted above all in the clashes between the opposing political players in the region, mainly the progressive and conservative forces represented in politics, government, and sectors of the media, as well as the political pressures placed on public communications by politicians from across the political spectrum, be it the Workers' Party (PT) or the Social Democrats of the PSDB. These have been the two main political players in the country that emerged in the aftermath of the dictatorship and have vied for power and control over the structures of government since then. Thus the highly politicized nature of the country's institutions and of the media still, in spite of the gradual growth of professionalism in newsrooms and the gradual expansion of the role of the media as a Fourth Estate, can be seen as being a key barrier and roadblock to wider democratization of the media as well as national development.

My research also investigated the nature of the medium of television and the different discourses and the similarities surrounding both the public and private broadcasting 'style', highlighting how the distinction between the two has become increasingly blurred and difficult to pin down. This is similar to what has happened with the commercial and public television channels in the UK, where the BBC has become in the last recent years more undistinguishable from commercial broadcasters like ITV in regards to programming and content. Many programs, genres, and discourses for instance are encountered in either one or the other, such as quality drama.

However, I have shown how there are still subtle differences between private and public broadcasting mainly in what we can classify as style, discourse and language, and types of approach to programming and content, including for instance the choice and selection of programs to occupy the peak slots. These differences are also perhaps more manifested in other areas than news, including in overall more subtle variations in the aesthetics adopted, the tone of the programming, and the selection of themes and topics covered by each station.

Commercial television stations like *TV Globo* for instance privilege during peak slots soap operas, news, or blockbuster entertainment whereas the public media has a tendency to include new broadcasts, historical programs, or quality drama. The peak slots of *TV Brasil* for instance were largely dedicated to journalistic, historical, popular culture programs, and/or documentaries. As both the interviews and the online survey that I conducted showed, the public media still does provide a wide space for the proliferation of debate, which according to many of the interviewees needs to be much better explored, as we shall see next.

SURVEY RESULTS

A persistent pattern that emerged from the answers of the online survey conducted with university students from UFRJ was that the penetration of public television is still small among most of them, and does not constitute the core part of their key television-viewing material. As stressed in some of the answers of the UFRJ survey, there is still widespread misunderstanding, lack of interest, and inadequate knowledge of what exactly the public media stands for and the place it should occupy in Brazil in the near future. The public media's potential and capacity thus still remains relatively unexplored. Many revealed, however, how they mainly watch commercial television, *TV Globo*, and cable and satellite television, with some abandoning more TV and shifting to the Internet instead, a trend that is happening worldwide and is significant in the UK largely among the younger generations.

These television-viewing patterns, which privilege commercial programming, largely serve to confirm the already known dominance of commercial television in everyday life in Brazil among most sectors of the population, especially the working classes but also across sectors of the elites. That said, a significant 71 percent of students of the UFRJ online survey said that they endorsed the public media. They recognized its importance, further underscoring the role that the public media could have in correcting market failure, complementing the commercial media, as well as contributing to wider media pluralism and democratization.

In terms of which television stations are most popular, most respondents of the UFRJ survey said that they watched *TV Globo* (97 respondents or 65 percent) and cable and satellite television (99 or 66 percent). Only 3 percent (4) chose the public media option and a slightly higher number opted for the Brazilian public station options, *TV Brasil* (8 or 5 percent) and *TV Cultura* (8 or 5 percent). These received similar percentages to the small open commercial television stations, *TV Record* (7 or 5 percent) and *Rede TV!* (4 or 3 percent). Channels *Bandeirantes* and *SBT* appeared in a middle position, with 25 or 17 percent for the former and 18 or 18 percent for the latter.

The responses for favorite TV programs were however quite varied. A popular TV choice was *TV Globo's Jornal Nacional* (38 or 25 percent). The option of the 8 o'clock soap opera appeared with 13 percent (20), although in the previous question concerning television genres, only 6 percent chose soaps. Nonetheless, the quantity of different programming selected is just another confirmation of how contemporary global media audiences have become much more fragmented than before. Forty-seven percent chose other programs that were not included in the list. The journalistic programs that appeared here as options were *Roda Viva* and *Observatorio da Imprensa*, which received respectively 1 percent each (1), as did the programs

Reporter Brasil, which is the main news broadcast from *TV Brasil*, as well as *Sem Censura*, the popular debate program previously broadcast on *TVE*, whereas Brazil's *Big Brother* scored 3 percent (or 4 answers).

Among the preferred programs freely listed by the respondents were films, popular national programs, or American series. Seven percent wrote 'films', whereas others chose the *TV Bandeirantes* program *CQC* (4 percent),[9] football (3 percent), *Friends* (2 percent), and *House* (2 percent). Other Brazilian programs selected included *Jornal das 10* (2 percent, *TV Globo* news program), *Jornal da Globo* (1 percent), and the popular long-running talk show *Programa do Jo* (1 percent). An interesting issue to observe was that the viewing of American series and programming has not transcended that of national ones. Programs such as *Jornal Nacional*, films, news, soaps, and football appeared alongside or above American series.

The UFRJ online survey results thus underscored how a segment of the audience in Brazil, as well as in the UK, still give significant importance to quality programming and are open to the correct combination of quality entertainment with in-depth information and debate. This came out quite clearly in the selection of options included in the survey, which appears in full in *Media and Politics in Latin America: Globalization, Democracy and Identity* (I. B. Tauris, 2011). Regarding the question on what attracted their attention to TV, the predominant answer was 'the quality of a program' (58 percent or 86). In second place was the option 'information' (22 percent or 33) chosen.

Thus such answers endorse the fact that television, be it in the UK or in Brazil, is expected by viewers to be both entertaining and informative, while at the same time also offering quality programming. These values are strongly associated with the public media ethos and indicate that journalists, producers, and other academics in Brazil have something to tap into if they seriously want to create a quality public media for the public interest, one that is capable of attracting a wide audience and of being influential in defining public policies and in serving as the country's core public sphere vehicle of debate.

Most audiences, however, still see little difference in terms of the *type* of information broadcast in news programs on either media, although the differences are very much more subtle, as mentioned previously. It is thus clear that both commercial *and* public TV are becoming increasingly blurred, and that there are many overlaps (e.g., broadcasting of news and drama in both) that are here to stay and that, on the other hand, should not be seen as serving to undermine the significance still of the public media platform in an increasingly changing and complex world.

The responses from the survey also detected a space for the production of high-quality drama and art films as well as music programming. This is still relatively ignored by the commercial media, or receives little financial support or incentive (e.g., Laurindo Lalo 2006). Thus the answers of most respondents made it clear that there is a significant space for the public media in Brazil in assisting in expanding debate, as well as investing in quality cultural and educational programming.

As the interviewees also emphasized in their choice of options in the survey, it is essential to develop a public media platform that is adequate for the needs of national citizens, be it in the UK or Brazil. Such a contestation casts doubts over the suitability of the application of the BBC model to many Latin American countries due to their historical, cultural, political, and social particularities. There is also the fact that in many, as we have seen, the relationship between political actors and the media is still marked by fever pitch tensions, a high degree of politicization still, and a long authoritarian and historical tradition of misuse of media structures for the personal interests of mainly individual oligarchic politicians, as well as media owners or private interests.

Regarding some of the key conclusions of my research, which I do not have sufficient space to go into detail here, it is important to note to start with that the public media platform in Latin American countries can really exist and contribute to strengthen press freedom *only if* it remains independent from both the public and the private sectors. It cannot fortify debate, serve as a vehicle for the public interest, or boost political pluralism, representing the whole of the political spectrum and the diverse interest of Brazilian society, as class liberal theory on the media would have it, *if* it continues to reinforce the tradition of use of the communication public structures for the personal interests of politicians or other vested private and/or commercial groups.

Thus the public media platform in Brazil needs to find its own formula of success, one that can go beyond the commercial fixation with audience numbers, which has tended to prevail in the public station in São Paulo, *TV Cultura* for instance, or the dependency on government support or on an officialdom editorial line, as *TV Brasil* has been accused of doing. Finally, the data collected in my research has largely revealed how, in spite of the challenges that it faces regarding political pressures and problems with lack of large audience numbers, the 'public' media in Brazil does still have a potential to be a force for change and expanding democratization, and contribute to quality debate.

TENTATIVE CONCLUSIONS

In spite of growing professionalism, liberal media cultures in the newsroom, and improvements in quality standards and balance criteria in the last decades due to market pressures, civil society demands, and political democratization (Matos 2008), the mainstream commercial media in Brazil are still highly vulnerable to both internal and external political as well as economic pressures. Given the political use still of the public communication structures in Brazil and in many other Latin American countries, the public media is thus also not immune from the negative impact of partisanship practices. Nonetheless, the contestation of this fact is no reason to dismiss its capacity to be directed towards the public interest, as the success of cases like the UK's PSB have proven for better or for worse.

Moreover, public media can also assist in internationalization and in better inserting Latin Americans in the global order. They can play a role in gradually reversing the historical legacy of political, cultural, and social marginalization imposed by the legacy of European (neo) colonialism. Therefore arguments in regards to the capacity of stronger public media in Brazil of being capable of serving as an instrument of media independence and freedom from both political and economic constraints (Matos 2008) are in tune with the times.

It does seem evident also that the philosophy and ethos of PSB has not died, and that various developing countries that are pursuing an agenda of massive investment in the public service platform are not going against the tide. These countries are pursuing a legitimate path of democratizing more knowledge by creating the means to strengthen public debate, to improve educational levels, and to invest in high-quality programming capable of boosting cultural emancipation and, in this way, slowly paving the way for wider cultural and educational equality and social integration of less privileged sectors of the population in the country's emerging public sphere.

NOTES

1. The report was the result of debates that were held at the 2006 International Intergovernmental Programme Council for the Development of Communications (IPDC).
2. The Communication Act of 2003 requires Ofcom to set quotas for UK national and international news as well as national and regional news on the commercial PSBs in both peak and off-peak viewing times. See the bibliography or information on the Ofcom reports.
3. See "Novas leis e projetos na America Latina esquentam polemica entre midia e governos" [New Laws and projects in Latin America heat polemic between media and governments], FNDC, September 29, 2010.
4. See FNDC interview.
5. There are 764 educational channels in the whole country, of which 459 are radio stations and 305 television channels. The other 'public' television channels in Brazil are TVE-RS, Parana Educativa, TV Cultura SC, TVE-ES, TVE Bahia, TV Ceara, Rede Minas, TV Brasil Central, TV Rio Grande do Norte, TV Cultura PH, and TV Palmas. The public sector platform and decision-making organ is composed also by the state radio station, Radiobras, Radio MEC, the Cabinet of the Presidency, and the Rio state television, TVE Brasil.
6. "Ipea sugere medidas para democratizar a midia no Pais" [Ipea suggests measures to democratise media in the country], Lara Haje, Camara dos Deputados, November 11, 2010.
7. Interviewed by telephone on August 5, 2010.

8. "Bernardo diz que discussao caminha para ter duas agencias na area de comunicacao" [Bernardo says that discussion is about having two communication agencies], FNDC, February 16, 2011.
9. CQC (*Custe of que Custar*), or *What It Takes*, is a program that mixes journalism with humor. The program consists of a group of reporters asking embarrassing questions to celebrities.

REFERENCES

Blaug, Ricardo, and J. Schwarzmantel, eds. 1988. *Democracy: A Reader.* Edinburgh: Edinburgh University Press.
Curran, James, and Myung-Jin Park. 2000. "Beyond Globalization Theory." In *De-Westernizing Media Studies*, edited by James Curran and Myung-Jin Park, 3–19. London: Routledge.
Dunleavy, Patrick and O'Leary, Brendan (1987) *Theories of the state: the politics of liberal democracy*, Handmills: Macmillan Education
Hallin, Daniel, and Paolo Mancini. 2004. *Comparing Media Systems: Three Models of Media and Politics.* Cambridge: Cambridge University Press.
Held, David. 1995. *Democracy and the Global Order.* Cambridge: Polity Press.
———. 2006. *Models of Democracy.* Cambridge: Polity Press.
Keane, John. 2000. "Structural Transformations of the Public Sphere." In *The Media, Journalism and Democracy*, edited by Margaret Scammell and Holli Semetko, 000–000. Hants., UK: Dartmouth.
Lalo Leal Filho, Laurindo. 2006. *A TV Sob Controle: A resposta da sociedade ao poder da Televisao.* São Paulo: Summus Editorial.
Lins da Silva, Carlos Eduardo. 1990. *O Adiantado da Hora: A Influência Americana Sobre o Jornalismo Brasileiro.* SP: Summus.
Matos, Carolina. 2008. *Journalism and Political Democracy in Brazil.* Lanham, MD: Lexington Books.
———. 2012. *Media and Politics in Latin America: Globalization, Democracy and Identity.* London: I. B. Tauris.
Moraes, Denis de. 2009. *A Batalha da Midia: Propostas e Politicas de Comunicacao na America Latina e Outros Ensaios.* Rio de Janeiro: Paes e Rosas Editora.
Norris, P. 2004. "Global Political Communications: Good Governance, Human Development and Mass Communication." In *Comparing Political Communication: Theories, Cases and Challenges*, edited by Frank Esser and Barbara Pfetsch, 115–51. New York: Cambridge University Press.
Possebon, Samuel. 2007. "O Mercado de comunicacoes: um retrato ate 2006." In *Politicas de comunicacao: buscas teoricas e praticas*, edited by Murilo Cesar Ramos and Suzy dos Santos, 277–305. SP: Editora Paulus.
Raboy, Marc, ed. 1996. *Public Broadcasting for the 21st Century.* Luton, UK: University of Luton Press.
Santos, Suzy dos, and Erico da Silveira. 2007. "Servico publico e interesse publico nas comunicacoes." In *Politicas de comunicacao: buscas teorias e praticas*, edited by Murilo Cesar Ramos and Suzy dos Santos, 49–83. São Paulo: Editora Paulus.
Scannell, Paddy. 1989. "Public Service Broadcasting and Modern Public Life." *Media, Culture & Society* 11: 135–66.
Schramm, W. 1964. *Mass Media and National Development: The Role of Information in the Developing Countries.* Stanford, CA: Stanford University Press, 1–58.

Straubhaar, Joseph D. 2001. "Brazil: The Role of the State in World TV." In *Media and Globalisation: Why the State Matters*, edited by Nancy Morris and Silvio Waisbord, 133–53. Oxford: Rowman & Littlefield.

———. 1996. "The Electronic Media in Brazil." In *Communication in Latin America: Journalism, Mass Media and Society*, edited by Richard R. Cole, 217–45. Wilmington, DE: Scholarly Resources.

Sparks, Colin. 2007. *Globalization, Development and the Mass Media*. London: Sage, 126–76.

Voltmer, Katrin, and Rudiger Schmitt-Beck. 2006. "New Democracies without Citizens?: Mass Media and Democratic Orientations—A Four Century Comparison." In *Mass Media and Political Communications in New Democracies*, edited by Katrin Voltmer, 199–211. London: Routledge.

Reports, Documents, and Articles

Ofcom's *Public Service Broadcasting Review 2008 Phase*. April 2008. Includes data from the 2007 Ipsos Mori PSB quantitative survey, Ipsos Mori PSB workshops, PSB Tracker, PSB quantitative study from 2003, 2004 PSB Review, and Ofcom's Media Literacy Audits.

Ofcom's *Second Public Service Broadcasting Review—Phase Two: Preparing for the Digital Future*. September 2008.

United Nations. 2010. *Media Development Indicators: A Framework for Assessing Media Development*. Paris: UNESCO.

United Nations Program for Development. 2004. *Democracy in Latin America: Towards a Democracy of Citizens*.

10 A Bahian Counterpoint of Sugar and Oil

Global Commodities, Global Identities?[1]

Livio Sansone

Sugar and oil are possibly the first and the second key global commodities. There have been, of course, other global commodities, such as salt, iron, cocoa, coffee, and cotton—and there exist a growing set of publications on the history and anthropology of global commodities—but, for a variety of reasons, their impact on identity formation and grand ethnic or national projects has been less intense. Moreover, sugar and oil as global commodities can be seen as paradigmatic of their epoch because in many ways they are a sign of their time and an icon of power: the universal language of sugar and its technology was Portuguese, sometimes together with Spanish. Sugar became a commodity that characterized and in many ways represented the Portuguese empire and the period of Iberian domination of the Atlantic. In the case of oil drilling and transformation into fuel, right from its start in the end of the nineteenth century, the technical language—after all, a global commodity jargon—was and is still is predominantly English and most of its technology has been thus far produced in the US and UK. Oil and the technology it enables come to represent the stage of modernity the global language of which has been English.

This chapter explores the effects of sugar and oil on identity formation or, more specifically, on the making of blackness and whiteness. It focuses on the region surrounding Salvador, Bahia, where both commodities have a great impact[2]—sugar, and its plantation society (Wagley 1960), starting from 1550 and oil from 1950. Oil drillings and, later, the construction of a very large oil refinery occur starting from the early 1950s in a region that had thus been dominated by sugar cane plantation and sugar mills. After comparing life under the domination of these two different commodities, I connect them to the issue of a transnational black identity created across the Atlantic, drawing from a common past of slavery and a more recent past and present where racial hierarchies still penalize populations defined as black. Lastly, I try to combine both the cultural hegemony that accompanies the economics of a global commodity and the influence of the Black Atlantic (Gilroy 1993) with a number of singularities that characterize this region of the state of Bahia. This part of Bahia is emblematic for other regions of Brazil and other countries in which oil exploitation comes to substitute other monocultures (such as sugarcane, cocoa, or small-scale fishing) while creating, often quite suddenly, a wholly different local economy, with new global connections, higher salaries, new conspicuous spending patterns, new values associated to certain forms of manual labor, and technical skills, and an altogether new form of evaluating what a good job is.

The research for this chapter is part of a larger project that combines both my present intellectual concerns: the history of African American studies in Bahia starting from the late 1030s, a period that culminates with the visit of Franklin Frazier, Lorenzo Turner, and Melville Herskovits to Bahia (1940–43) (Sansone 2012); and the development in the region around Salvador da Bahia over a long period of time of what I like to call a culture of inequality—the naturalization of difference (Tilly 1998) that makes acceptable or bearable life in a context of excruciating inequality. It is a kind of social and cultural pact between the haves and the have-nots. This culture develops in time and takes time to recede. Of course, as much as continuities, I also try to elicit ruptures in the experience of this rather eschewed social pact. This Bahian pact follows a number of local rules, but also shows similarieties with other such pacts in the Global South with a history of durable inequalities especially in other parts of Latin America and Africa—after all the region and the continent with the highest Gini index of inequality.

The study of the persistence of durable and extreme inequalities, as well as the specific cultural forms and social strategies inequalities help to create, can gain new insights by focusing on the long history of specific regions, identified as open-ended and yet territorialized systems of opportunities. In this respect highlighting the situation of this specific region of Brazil can help identify how such inequalities are constructed, kept in place, and manage somehow to reproduce themselves across time and generations. Some regions can be especially crucial, such as those that experience a rather sudden transformation from a monoculture to a 'mono-industry'. The region around the Bahian Municipality of São Francisco do Conde, ca. 25,000 inhabitants in the year 2000, part of the region known as the Recôncavo of the Brazilian state of Bahia, 80 kilometers from Salvador, is a case in question: it is interesting for its past as one of the cradles of the sugar plantation society in Brazil as well as for its present on account of its very large per capita revenue from oil refinery and transformation that is combined with extremely high Gini index, which measures inequality.[3]

This research is based on fieldwork among two different though sometimes interrelated groups: former workers of sugar mills and their offspring; and the first-generation oil workers and their offspring. To this we added, of course, material from archives and from accidents—such as finding the log book of the Dao João plantation and sugar mill that has taken a central place in our research. In order to be able to describe the long period of time from 1950 to the present, our research focuses on two age groups, the older generation, now in the age bracket of sixty to ninety years, and the young generation, now in the age bracket of fifteen to thirty years. From January 2007 to January 2009, after two years' research in archives, oral history, in-depth interviews, and participant observation, our team (consisting of myself and four senior undergraduate students) applied a survey in a representative sample of 500 families, distributed in the several districts of the municipality. The survey, which focuses on the perception of inequality in relation to consumption, racial terminology, leisure, and work/unemployment, will be analyzed elsewhere. This chapter relates to the qualitative part of our research, focusing on one specific plantation and sugar mill and later one region for oil extraction, rather than plantation life more generally in Bahia or the oil industry as such in Brazil.[4]

My research is in many ways a follow-up on the great Columbia/UNESCO project on race relations in Brazil (Chor Maio 1999; Pereira and Sansone 2007), which carried out fieldwork on plantation society in 1950–53 in the same region (Wagley 1963; Wagley and Roxo 1970) or in other parts of the State of Bahia (Harris 1958). Those were the years when oil drillings started in the region and plantation life was seen as the epitome of backwardness (*atraso*) in the state of Bahia. In fact just opposite to the Dao João sugar mill and plantation studied by William Hutchinson (1957) and later Maxine Margolis (1975), the then recently founded National Oil Company, later renamed Petrobrás, built the first sizeable oil well camp in Bahia and named, in what was then perceived as a cultural provocation against the ruling plantation system, with the same name of the sugar mill—it became the Dao João Oil Camp. Of course, in those early years of blunt cultural and economic opposition of the world of oil drilling and refining to the world of sugar and alcohol production, nobody, really, could have imagined that through ethanol, sugar cane would have become an important part of car fuel in Brazil only twenty years later, already from the mid-1970s.

Besides comparing social and race relations and hierarchies in the age of sugar and oil, I also investigate the different systems of memory that sugar and oil have developed in that region. As I show later, the oil industry has had a great impact on the system of memory and remembering. Here I am concerned with the consequences of the arrival and development of the oil industry first and the revenue of oil royalties later for family life, identity formation, religious life and notions of blackness.

In many ways, the study of the becoming of durable and extreme inequalities is a study of the different stages of modernity and its consequences for the system of domination and social hierarchy as well as for the kind of resistance and set of expectations it leads to. In my project I and my assistants isolated, for analytical purposes, three stages in the construction of inequalities, each of which is characterized by a main driving force in economy:

- A first period in which the sugar cane and its memory system and culture determines local economics as much as its transnational connections—leading actors are important and well-known local

families of capitalists with constant shortages of capital—capitalists
without capital;

- A second period in which oil and its much more powerful memory
  system and culture becomes quite suddenly and for a few decades the
  driving economic force—the single leading actor is (faceless) capitalism without capitalists;
- A third period characterized by oil-related revenues for the municipality of S. F. do Conde that make possible an oil-fueled populism—
  the leading actor is wealth without social contract, such as in most
  hydro-carbureted societies (Karl 1997; Coronil 1997).

SUGAR AS ICON OF THE PAST

In Brazilian traditional historiography, mostly though not exclusively in the
case of popular historiography, sugar and alcohol production, especially in
the northeast of the country, is seem as representing the past.[5] In this tradition sugar becomes an icon of *atraso*—an economic (under or sub-) development characterized by the intrinsic absence of technology, innovation,
and modernity as well as by labor relations distinguished from 'modern'
labor relations for being centered on status (and hierarchy) rather than on
contract—'patriarchal' relations, as they are defined in this historiography
tradition and in a literature genre that made of the *engenho* and its plantation the core of its narrative and of which in Brazil José Lins Rego was the
main and most widely read interpreter. In fact sugar made the first globalization possible: it had a global market and technology and, when associated with slavery, it created pretty similar working and living conditions
in different regions of the world. It has had, historically, a homogenizing
effect on labor relations, technology, and the world of finance and credit. It
has had, historically, a homogenizing effect, also on food taste! As Sidney
Mintz (1985) brilliantly demonstrated that in order for sugar to become
an authentically global commodity also a global taste for sugar had to be
created—after all eating sugar in our modern doses is not a 'natural' thing.
It was only when the British working class started to have a diet heavily
relying on sugar (in marmalade, teas, cakes, etc.) that the demand for cane
sugar stabilized and grew ever since at least until the arrival of beetroot
sugar in Europe. Over a long period of time and well into the nineteenth
and twentieth centuries also in Bahia the most advanced technology of
its time was in the sugar mills (Schwarz 1976), and it created both capital
accumulation and the proletarian condition.

In interviewing the now aging former workers of the Dao João mills,
we came across a specific blend of working-class culture, a key element of
which was a constant yearning for land, freedom of moving, house ownership, time for your own, respect (men) and reputation (women), and a
disposable income to be spent on the body (clothes, soap, hair care, etc.).

The old former workers remember of the mill and even more so of cane cutting with a mixture of nostalgia and fear. They recount of the constant paucity of food and of how they had to look for extra food in the week-ends and after work. The company gave no plots of land for the worker to have an orchard (the managers we interviewed stated, to the contrary, that all workers were entitled to a plot and that many grew tobacco, mostly for their own use). The nearby mangrove was the source of most extra proteins, yielding crabs, shells, and some fish. The mangrove belonged to nobody or, rather, was seen as belonging to everybody.

Also the attitudes towards property remind us of working-class culture elsewhere: the contested meaning of taking from the land and the mill (pillaging versus reappropriation; poaching, fishing, and 'keeping' or 'nicking' charcoal and molasses considered by the workers as acquired rights and perks adding up to your salary and by the masters as signs of intrinsic lack of discipline and inclination to theft). Time was established by the mill siren, possibly one of the few clocks of the area, always in combination with tidal time—water transport, fishing, and harvesting depended on the moon and tides. When remembering the mill and the plantation, the memory of former workers in the cane is a bittersweet mixture: there was working-class solidarity and community, but also scarcity, hunger, bad health, sick and dying children and little option altogether.

While reassessing that body practices are part and parcel of working-class culture, our research has been trying to capture how people felt about the body, beauty, and fashion in those days. We heard that even in the constant paucity of fabric characterizing daily life, workers insisted on dressing nice and clean at weekends. After work everybody immediately washed and changed clothes. After work one simply tried to think of something else than the hot steam of the mill or the scorching sun of the plantation. Memories are of sweat and heat. Cleanness seemed to have been a way to regain humanity and maintain a reasonable living standard in the small and crowded houses that dotted the plantation and where the workers were 'allowed to stay' (without ever becoming owners). Cleanliness (together with an emphasis on orderly family life in spite of poverty) was also a way of keeping a distance from the large crowds of extra sugar cane cutters. These were hired just before the harvest to buttress production and usually came from the drier inland. In the recollections of the former workers we interviewed in SFC, these seasonal workers were often represented as being a combination of very hardworking people who were paid per production, tough or even violent men, simpletons and strikebreakers that the masters used and put up against the 'regular' local workers who lived on the plantation all year around. Cleanliness was pleasant, but was also a way to mark a position—close to city life and what one perceived as being *moderno* (in popular Portuguese of Bahia, this word means both 'modern' and 'young'). A piece of soap was a welcome traditional present to a newborn kid or a recently married couple.[6] Fashion mattered too. Being abreast of fashion gave you quite some status among your fellow workers. Metropolitan fashion was mediated by the local seamstresses, who got their tailor models (*modelitos*) from the occasional magazine that one of the several local women who was a housemaid in Salvador brought on one of the periodical trips back home (traveling to Salvador was one day by boat—it is now ninety minutes by bus). On some occasions dresses were tailored after a dress that an upper-middle-class woman had donated to her housemaid. Men knew about fashion and trends from the several colleagues on the mill or the plantations who traveled to Salvador for work —such as the sailors on the barges carrying sugar and rum.

Workers in the mill had their cultural life and leisure activities. Samba (especially the local version of *samba de roda*), capoeira, and a set of religious rituals combining popular Iberian Catholicism with Africa-derived rituals were established elements of social life, and especially from the 1950s samba and capoeira started to be performed) also in the large court inside the mill before the house of the family of the owner—on special Saturdays even the daughter of the owner had to join the samba and show how good she was at it. In a similar fashion, the two well-known 'priests' of what now would be called umbanda and candomble houses were respected, and the owner of the mill would lift his hat when passing in front of one of these houses. Interestingly, as we were told, the same daughter who had to samba in the courtyard of the mill for the workers to admire her was not allowed to join any samba dance in Salvador.

In fact the Dao João mill, with its approximately 1,000 workers plus one more thousand in the cane fields, was the fulcrum of modernity in the region: from the 1940s to 1969 when it went bankrupt, the mill had the largest grocery store and the only cinema in the surroundings. In the weekends people came from the surrounding areas to the small village just opposite the main entrance of the mill for partying, listening to music, buying clothes and fabric, as well as just getting to know what the news was. It had a special railway, a port, and a fleet of barges, the only trucks in the municipality while simply concentrating about every skilled worker of the region.

My argument is that any intrinsic technological, political, and cultural backwardness associated with the sugar cane industry cannot be taken at face value and has in fact to be understood as a cultural construction. As from the 1970s when industrialization around the development of the largest petrochemical plant of Latin America located just twenty miles from S. F. do Conde boomed, both the haves and the have-nots, for obviously opposite reasons, had to represent the sugar plantation as a reminder of the past, rather than the mother of much of the present. The aim of such representation of the world of sugar was to prevent any material and symbolic claims based on slavery or the slave-master condition. As a matter of fact, soon after the mill went bankrupt in 1969 most of the workers' rows of small houses were rapidly bulldozed and the machines of the mill sold to another mill. The past had to be wiped out. Only a few of these houses, sitting right in front of the entrance of the mill, were left standing and just because some of the most militant or resilient workers lived there—it is among them that we did most interviewing. As I explored elsewhere slavery cannot be remembered that intensively, nor can it be transformed easily into heritage (sites) when its memory is still alive in both popular and elite culture (Sansone 2002b).

OIL AS ICON OF MODERNITY

Opposite to sugar, in people's memory official publication and academic literature on economic development in Brazil,[7] oil often represents modernity. In more recent times oil has also come to represent a new stage in capitalism characterized by the parallel growth of wealth and inequality. My argument is more complex and posits that in Bahia oil—through the state company Petrobrás—has made the transition to full-fledged modernity a lot easier, but was not the great leveler it had announced to be in the 1950s (de Azevedo Brandão 1998; Costa Pinto 1958). Oil has caused a set of changes, but has corroborated other tendencies. Here let me suffice by listing a set of changes as they are presented in the interviews with former workers:

1. For the first time in the region, technical skills are highly valued in the labor market—skilled workers and technicians that had learned their skills in the mills, as apprentices, are lured by the oil industry with much higher salaries combined with a less hierarchal shop floor culture. To an extent, also heavy manual work is given a higher status because in its first decade the state oil company hires also thousands of unskilled laborers from the region to build roads, ports and the refinery.

2. Opposite to the sugar mill and plantation workers, those employed by Petrobrás after work like to show their blue overall and metal-toe shoes dirty with mud and oil. We hear a lot of stories of oil workers coming off the pier, where oil company boats deposited every evening those working on one of the several oil wells in the bay just one or two miles off shore, and walking straight into a bar, with dirty clothes and wearing their yellow crash helmet. The story goes that those workers would have then bought several rounds of drinks for every customer in the establishment. Mud and oil had to be shown and even had to be, as it were, acted out as an act of revenge against the haves. Conspicuous consumption was often the way to convey the message to the traditional sugar industry-based town elite. So, we were told that one of those workers coming off the pier offered twice the price for a fish for sale on the local market just for the pleasure of taking it off the

hands of the mayor, a representative of the sugar industry elite, who, as an act of public generosity, had ordered that fish to give it as a present to a poor and sick old lady. Showing off your hardworking body as well as the money that you had earned through it was the message the new proletarian elite sent to the old political and economic elite (until 1972 all mayors had come from a few families of sugar mill and plantation owners).

3. Petrobrás changed the structure of employment radically also in terms of gender, by hiring until a few years ago exclusively males, especially for the actual drilling and the outshore platforms (see Farias 2003). In the sugar mill and plantation, women participated in the production process, at least in the busiest months of the harvest and grinding of the cane. Petrobrás institutionalized, for the first time in the lower class, the role of the housewife and, if the husband died, which occurred all too frequent especially in the first twenty years because of the very high incident rate, the role of the *pensionista*—a housewife receiving the retirement benefit of a deceased husband. In the same line came general retirement rights (mostly unheard of in the sugar industry), health care for the whole family of the worker, and literacy and technical courses for the workers (after the 1980s Petrobrás hired increasingly only skilled workers and candidates with technical degrees). These provisions went together with a complex double process as regards family life: one the one hand, both the company and wives as well as unmarried mothers pressured towards formalization of fatherhood and the social benefits that came with it, which strengthened nuclear family ties; on the other hand, this process of formal recognition transformed what would have otherwise been single mothers into receivers of aliments and therefore forming a second and sometimes a third family for the oil worker.

4. Good health care for the workers and their families, something Petrobrás was proud of, means, as an oil worker widow told us, children stopped dying. In a very short period the health condition of these people improved dramatically. It is worth stressing that this is the positive aspect of the golden years of Petrobrás that women tend to remember the best. Men, on their account, like to remember the new opportunities for learning skills and consuming. In Mara Viveiros's (2002) terms, oil industry men are recalled as *quebradores* and *cumplidores*—they are the best providers available in the pool of marriageable men, but also the most extravagant, streetwise, and promiscuous men in the region.

5. With a disposable income comes home ownership—as opposed to living in house on the mill's land for which no formal rent is paid, but fidelity to the company is required (*morar de favor*).

6. Promotion of literacy powerfully affects the mechanisms of memory and notions of rights.

7. Formal and equalizing, as opposed to caste-like work relations. Trade unionism is possible and, at times, even stimulated by Petrobrás—to be discouraged again during the military dictatorship of 1964–83.

8. Disposable income—the 'fridge generation', as was known the first-generation workers who were able to afford a fridge, uses conspicuous consumption to gain access to visible forms of power enforcement.

9. These economic and social changes are accompanied by a process of diversification, segmentation, and specialization in the domain of religious experience. From the fifties to the nineties there is trend from a situation of monopoly on the part of the Catholic Church, in association with popular Iberian Catholicism and a set of relatively informal Afro-Catholic tradition often having their rituals performed on the plantation rather than in town, to a situation characterized by what sociologists define as a religious market, consisting of the Catholic Church, popular Catholicism, 'properly' established Candomble houses, and a variety of Pentecostal churches. There is some evidence that the 1950s were the years in which two important steps were taken, both of which were performed by families of Petrobrás workers: the founding of the first chapter of the Pentecostal church Assembléia de Deus, and the establishment of the first two candomble houses organized according to a model that was largely inspired by the main 'traditional' houses based in Salvador. As regards religiosity of African-Catholic origin, in those years one sees a transition from informal and often movable cult places to temple houses, as well as from practices that were often defined as witchcraft (*bruxaria*), also by those who hold it in high esteem, to what is being called religion or just candomble.

10. Interestingly, in the interviews with retired oil workers, the expression *negro*—which has nowadays become the term to define people of African descent and that are proud of it —appears in their narratives only as from the years after the settlement of Petrobrás in the municipality. This is something we are still exploring, but it is clear that a certain degree of black pride comes with both oil industry trade unionism and some of the symbols associated with the oil industry itself. One perceives easily how black oil workers are proud of calling the 'mineral' *ouro negro* (black gold) or how easily they make the pun of Petrobrás & Pretobrás (preto is the color term for black).

Over approximately the last two decades, a new economic and social context is taking shape in the micro-region through a combination of increasing oil revenues, an overrich municipality that sustains a relative small local elite with high salaries and special benefits on the one hand, and lots of poor people that depend largely on the municipality for work, social benefits or welfare, and favors on the other hand. One of the reasons for choosing SFC for a research on extreme inequality is that the municipality is the first or the second in Brazil in terms of wealth per head of the population (26,500 inhabitants in 2000) while also being a champion in terms of a poor Index of Human Development. The extremely high and growing revenues from royalties, because of what is established in the progressive postdictatorship Constitution of 1988 with its emphasis on decentralization, are mostly retained in the municipality where the oil well or refinery is rather than been appropriated by the federal government as in the past. These large sums of money are managed by a relatively small number of people who run the town hall. In a few words, a new local elite becomes amazingly empowered with such royalties and has come to represent a third power in the history of SFC, possibly the most powerful ever, after the visible sugar barons (capitalists with little capital) and the invisible chiefs of Petrobrás (capitalism without capitalists). A new and more recent strand of inequality is added to the traditional one (Comaroff 2001).

Such situation, of a royalty-driven economics and elite, comes into effect powerfully as from the late 1980s, the period of redemocratization, and in which Petrobrás suspends oil drilling and the exploration of wells in SFC (many wells were reopened in 2006, stirring up a renewed interest among young people for a job in the oil industry) and limits its use of the territory to the large oil refinery—which yields most of the above-mentioned royalties.

In spite of this context, determined by renewed inequality, ravaging corruption scandals, and a set of impeached mayors (mostly allied with conservative political interest groups in the State of Bahia), SFC was the municipality in Brazil that produced the highest percentage of votes for Lula's presidency in 2002—93 percent!

MEMORY

As said, sugar and oil are associated with different infrastructures—or regimes—for memory. The world of sugar produces three sets of memory: from below, from top down, and from the point of view of the Communist Party and its spokespersons. The first set tends to be individual centered, if not individualistic. It is the expression of the illiterate, mostly proletarians *an sich*, with a class consciousness that we have to find in between the lines: a conglomerate of personal accounts mostly of a dyadic relationship with a charge hand or an administrator. In this group the memory of slavery and labor on the plantation and the mill can be detected, at times, in the narrative and text they produce in samba lyrics, tales, proverbs, naming of people and places, prayers, and the way of celebrating the festivities of certain Catholic saints such as S. Anthony and S. Roque. There is also a lot of silencing. Even when one hears of humiliation or resistance, it is mostly told in terms of, for instance, an individual, even violent reaction to a personal offense—such as being shouted at by a foreman or administrator in the presence of fellow workers. The second set corresponds to the haves: it consists of a series of rather sweetened memories that are pretty well structured through family albums, family genealogy trees, local historians' and anthropologists' publications (some by self-taught ethnographers or historians), several autobiographies, or nostalgic-tinged novels 'about the past'. The third set corresponds to memories of collective resistance that can be found in the articles of *Momento*, the Bahian Communist weekly newspaper quite popular among sugar cane cutters and mill workers published between the late 1950s and the early 1960s: in the interviews with workers or in statements by workers, the plural we/us is the only form. In the *Momento* these laborers are always addressed in the plural, unless the article dwells on some pitiful cases of ill-treated workers. The will of the workers, it is suggested, is positive when expressed in the plural.

If one goes back to the in-depth interviews, we see that much of the sugar industry workers' resistance is the result of individual attempts and claims—mostly attempts to regain humanity by achieving 'respect' and even individuality. When workers in the mill and the fields mobilize a collective identity this is mostly masculinity—*hombridade, de hombro a hombro* (shoulder to shoulder), as ethnographer Camara Cascudo said already many years ago. Masculinity is the link between the mill owner and the skilled worker, and between the foreman and the sugar cane cutter. Infringing on the silent rules of respect, such as shouting at another man, not to mention threatening another man with violence, especially in public, can be conducive to (violent) reaction. Of course, this bears some reflection on the role of honor in claiming (male) identity as well as the *persona* in a society marked by slavery.

Color is part of the workers' narratives only in certain, rare, episodes. The proletarian condition, in most cases, is seen as less encapsulating and limiting than the one of black person—in the sugar cane fields as well as today in the oil plants. Color tends to come up when we as interviewer stimulate the topic, but not spontaneously. In the younger group, which tends to be altogether better educated and more often unemployed than their parents, the term *negro* is used more often and there is a slightly more pronounced inclination in recognizing racial discrimination as a fact of Brazilian society—possibly, as I argue elsewhere (Sansone 2003a), this results from a more 'mixed' social life and the greater chance of maneuvering across different social and color groups when compared to the parents, who tended to be more 'local' and are much less mobile in their leisure time in public. It must be stressed that the term *negro* has changed value over the last century, in this region and in Brazil as a whole, from a term perceived as an imposition (something other people would call you like) or a pejorative to an assertive term now seen as part of self-definition. It seems that blackness becomes interesting, as a factor conducive to a higher self-esteem, only when it can be perceived, at least to a certain extent, as a choice rather than as an imposition.

The mechanisms of memory are among Petrobrás workers somewhat the opposite of those among the workers in the cane fields and mill. Literacy, trade union press, Sindipetro (a powerful and influential trade union), national advertising campaigns, company bulletins, the sheer existence of very conspicuous oil plants (as opposed to the ruins of the Dao João mill), and over the last few years, even a project, inspired and led by the national management, to recover the history and memory of Petrobrás—the Projeto Memória tries to turn a corporate culture into (national) heritage.

No wonder that in the region it has been a lot easier to gather material on the last fifty years, dominated by Petrobrás, than on the much longer period before that was dominated by the production of sugar and alcohol.

FROM POPULAR CULTURE TO AFRO-BAHIAN CULTURE

Culture change has been part and parcel of our research. Let us see when, how, and why Africa and its trope or color terminology enter the above-mentioned structures of memory and the realm of cultural production or the narratives about culture. After all the Black Atlantic—the feeling of a shared past and present for population of descent in different countries and regions across the North and South Atlantic Ocean—exists because a set of common icons are remembered and reworked across different regions: Africa, of course, but also 'race', notions of beauty, soul, rhythm, and sufferance/resilience (the collective memory of great injustice).

As we have seen form other researchers researching the decades immediately after abolition (see Mattos 2005; Fraga 2007), in the memory of our informants, some of which date back to the 1920s and 1930s, the language of color and race was avoided, for different reasons, by both workers and bosses. The acceptable language between different social groups was that of class (worker versus charge hands, administrators, and owners) or gender. For example, former workers remember with some pleasure that in those days one could talk to the owner Dr. Vincente 'man to man'—although one was mounting a horse and the other was holding a machete. Dr. Vincente was known to be tough but fair, often adding a little bit of money to the pay envelope of certain workers while insisting that the amount was the official (and minimal) wages for everybody.

To begin with the place of Africa in popular culture and narratives of blackness changes: from implicit in the age of sugar to explicit in the age of oil. The foundation of the two most important candomble houses in the fifties, and more strongly from the 1970s, play a key role in starting again to remember and quote Africa in SFC. In fact what has been called the re-Africanization of Bahia is a process that started in the city of Salvador and that later bounced back on the rest of the state of Bahia. Curiously, even though the Recôncavo rediscovers the trope Africa largely after a Salvador-centered model, this region represents an area that is by many identified as being the roots of many of the aspects composing what was defined, from the thirties onwards, as *cultura afro-bahiana*. The Recôncavo is home to the ingredients for Afro-Bahian food, to clothes, percussions, *samba de roda*, ship and canoe building, and '*bruxaria*'—that is where the powerful herbs and their magical power come from. Anyway, in SFC, as in other urban areas in the northeast of Brazil, one sees that in order to gain the acceptance of the local elite, and become part of the cultural milieu of the municipality, candomble houses have to resemble in structure, liturgy, and even name an ideal model represented by a set of 'authentic' candomble houses in Salvador, mostly associated with the Nago/Yoruba nation. Several smaller and poorer candomble houses are ignored by local politicians, and its rather generous though unforeseeable system of support to cultural and religious groups, because they do not fit that ideal Salvador-inspired model. It is worth stressing that few of these houses ever join the official federation of Afro-Brazilian cults and that the two houses I mention have in their certificate that they are of the Angola nation even though they have both recently adopted Yoruba names.

The process of transformation of cultural forms and artifacts that were once not named at all (but simply done or performed customarily) into 'popular culture' as from the 1970s and into 'Afro-Bahian culture' as from the 1990s has its own agents (see Vale de Almeida 1999). I investigate local and external actors, agendas and agents in such cultural revolution. In the 1990s the municipality, able to pay better salaries than Salvador, attracted scores of schoolteachers and cultural producers from Salvador. Several of them were black activists or cultural entrepreneurs. SFC became well known all over Bahia for its large open-air concerts, S. Johns feasts, and carnival—very large events for such a relatively small town. The mayors and their staff started to be convinced that investing in culture was worth the while, that cultural creativity would have put SFC in the picture of tourism, and that culture, now more and more Afro-Bahian culture, was something to promote. In 2005 on the verge of local elections, the municipality distributed thousands of free T-shirts announcing a concise but poignant text: S. Francisco do Conde Culture Capital.

This overall change in the field of culture and identity relates of course to important changes in the intersection of color, body, beauty, and gender. Brazil is a country that has made of the mixing of races and the making of phenotypical variety something to be celebrated in popular culture while also establishing over time a complex pigmentocracry—with the pure white and pure African at the extremes. Also in the Recôncavo such scale of classification does not hinge mostly on color proper, but on a combination of skin color, hair type, thickness of lips, nose and head shape, and type of feet (people swear to me that some black people really have African feet, large, sturdy, and with a flat heel). The *qualidade* (literally, quality) of a person is a result of a combination of these traits with the signs on our body of manual or dangerous work, such as calluses, broken or dirty nails, scars, and skin diseases (mostly fungi). Reading the social position of a person in the body in such a way is not an easy thing, and being able to do so makes you the ultimate Bahian. Fluency with these codes is, however, imperative in doing fieldwork because asking too straightforwardly about racial discrimination can inhibit an informant. On the contrary, asking about the ideal husband, good hair, what is a beautiful body, or just love (which conjures up all these elements) has worked for me as an emic trigger—these are the kind of questions that people like to answer and that strike a familiar chord among most informants because this is the way people talk about 'race' in Brazil. Of course, phenotype can be rather important in Brazil, where blond hair and blue eyes have long been associated with wealth and even modernity. In SFC, where according to the 2010 Census data white people are a mere 7 percent, most 'whites' are so by definition and, more than in the labor market or in contacts with the police, as in parts of Brazil where there are more white people, color/phenotype matter in the domain of courting and getting married. This is of course a system that has led to the creation of a racial habitus that is both specific to the region and transforms and reinterprets images of beauty, 'race', whiteness, and taste that can also come from far away. It is not a static and self-contained system. Some major changes took place with the advent of oil, as the blue overall case mentioned before whereby for the first time the dirty body of the (black) working man can mean status, whereas other changes have occurred over the past two decades on account of the (late) demographic revolution and the becoming of the 'young person/generation' as a new social group and the popularization of the notion of 'staying young and thus beautiful' now also in the lower classes. A further factor for change is what one can call the popularization of feminism that has affect the construction of the pool of marriageable men—men are now increasingly important not just as provider but as 'partner'. Nowadays these changes have affected the perception of citizenship, something that has now become also, in many ways, aestheticized. Being a satisfied citizen is also living a life in a healthy body that can be perceived as beautiful and experiences pleasure.

PLACING *NEGRITUDE* IN CONTEXT

The main specificity of S. Francisco do Conde, when compared to other sugar lands (Bosma, Giusti-Croder, and Knight 2007) or other regions where oil comes to dominate the local economy, especially after centuries in which another monoculture (the sugar plantation) had dominated (Karl 1997), is in the domain of culture, religion, and the language of race relations or racial hierarchies. It is in this domain that the mostly white elites have deployed strategies to maintain their position over a long period of time in a situation where, from the abolition of slavery in 1888 to the present, 'white people' have been very few and decreasing in number in the municipality.

However, one sees also 'global traits' in the domain of culture and identity at work in this region of Bahia, such as in the kind of black (youth) culture that is being created that revolves increasingly around the aestheticization of blackness, associated with body practices and politics, and less so around religious life (Sansone 2003a). To the contrary, religious life constituted the base of what has been called, especially in Salvador, Afro-Bahian culture. This is a culture centered on the practice of candomble and its cultural whole (consisting of samba, African-derived cuisine, special dresses and attires). Up until approximately fifteen years ago in SFC, candomble houses and their communities were, basically, the sole sites and media through which memory of Africa and slavery was preserved through complex genealogies and sets of 'local' tradition centered on 'religious families'—in a very hierarchical way doing the mediation between the present and a magic African past. Nonetheless, in spite of the fact that local forms of black cultural production and performance of blackness, as well as the ways through which the young generations express their dissatisfaction of social inequalities, use to an increasing extent the icons and language of what one could call global blackness or global black culture, there are certain aspects of life that seem to show a high degree of resilience to global cultural flows, even if they are wrapped in the appealing and seemingly universal language of *negritude*. Such specificities need to be highlighted in that they help us understand how local 'culture' functions as the lens through which people tend to read global flows. Although not detailed in this chapter, I can say that the language of conflict and negotiation, even of young people, is still imbued in the relatively local tradition hinging upon acceptance of extreme social inequalities and venting one's dissatisfaction indirectly and avoiding to create friction beyond the point of repair with the haves—be it the sugar mill owner, the Petrobrás chiefs or, nowadays, the mayor and his immediate staff.

In ethnic studies there has often been a conflict between cultural- and structural-centered explanations as regards the strategies, for example on the labor market, of ethno-racial minorities. Generally speaking, left-oriented scholars opted for structure, whereas right-wing-leaning scholars opted for culture. This is, of course, no way of getting by with the questions raised by the complex relationship between economic change, social mobility, and the ethno-racial position. In my research I have attempted to, as it were, escape such dilemma by focusing on the long-term formation of the context for today's interethnic relations. I seek to avoid the pitfalls of presentism that come with blunt rational choice theory approaches to ethnic identity formation .[8] In doing so we have to keep looking for both continuities and ruptures. In many ways '(ethnic) identity', as we say today, is a creation of modernity that can take shape only where and when the conditions for modernity and modernization are given (Gleason 1983). Conditions for modernity can exist, of course, even in the context of segmented and unequal access to the icons of modernity as well as to what is defined as full-fledged citizenship, as seems to be the case for Latin American modernity.[9] A good case is the relationship between negritude and modernity, a link made intimate by W. E. B. Du Bois and, later, Paul Gilroy. On the one hand one can say that modernity at whatever stage has always created condition for identity formation—and the redefinition of past allegiances. On the other hand, we have to take care to use today's interpretation of identity formation for analyzing past forms. Identities before the birth of identity (Hobsbawm 1997), and its canonization in the social sciences (Brubaker and Cooper 2000), were often voiced in very different ways—as 'culture', 'race', *campesino* instead of *indio*, African instead of black or *negro*.

My emphasis on the locality of identity formation does not, of course, underscore the fact that obvious similarities, continuities, and exchanges in black versus white identity formation can be detected across the Atlantic World. Antinational notions such as Atlantic World (System), African or Black Diaspora, and more recently and accurately the Black Atlantic have been a method for the reading of antiracism and struggle for emancipation as transnational rather than national phenomena, which communicate across nations, color lines, and class. The Black Atlantic, as method, however, is often presented in rather unilateral fashion—as a unique solution or as the sole source or method for explaining transnational similarities. I tend to believe that the theme of the Black Atlantic works hand in glove with other factors and is often intertwined with other powerful collective representations and systems of memory. In the case of the Bahian Recôncavo, the Black Atlantic has to share its influences with the following:

1. The Portuguese colonial style and the transnational networks it created.
2. Catholicism, in its high-brow and popular versions, which produced a special Brazilian sort of Baroque Catholicism—with a special emphasis on images as icons of both holiness and humanity, as opposed to what one could call the cult of the written word that grew with Bible-centered Protestantism.
3. The melodic tradition in music that in Brazil combines with what is generally considered as the African influences in music making (percussions and call-and-response have been often seen as key elements of 'Africanism' in music; Lomax 1970).
4. The economics, labor relations, and cultures made possible by the existence of global commodities—sugar and later oil. Each of these commodities developed through a certain ruthlessness as regards 'local' cultures and mores, by introducing global standards of quality, taste (in the case of sugar), price and technique.
5. Last but not least one has the universal experience and culture engendered by the class condition—both the working class and the upper class. In other words, the culture of the elites, for example on the sugar plantations, showed a bunch of common traits (e.g., how to dress and

talk, what to read, how to treat slaves and servants, politeness, attitudes towards technique and manual skills) right from the inception of the global circuit about sugar and cane alcohol. Also, slaves and, after abolition, free laborers in the fields and mills responded to the challenges of their working conditions in fashions that have always been extraordinarily universal—suggesting that, after all, the proletarian condition can produce a culture of work or waged labor that tends to be universal.

The main contention of my chapter is that the processes of identity formation are not ahistorical and neither are they inherently translocal (Handler 1994) even when they are subjected to the cultural hegemony of a global commodity such as sugar and oil. We have to be cautious in using the perspective of the Black Atlantic at all times, under all circumstances, and very often as the only method for explaining or even just representing traits in popular culture among people of (part) African descent in the Americas. In fact identities, and even those relating to the 'great identities' or transnational ethnic projects,[10] although using icons that have always been translocal, such as Africa, black, or white, are often related to specific localities and contexts. Even if one considers only parts of the world where living conditions have long been dictated by the economics of global commodities such as sugar and oil, there is *no* international identity game with universally valid rules. The making of identities, of course especially of 'great identities', is, however, always a case of transit between the global and the local as well as popular and high-brow uses of indigenous categories, the native and the analytical.

THE BROADER QUESTION

The last few years in our part of Bahia, but also in Brazil generally, have been a period of great and epochal change—especially crucial if one thinks that Brazil had long been considered a country of very limited opportunities for social mobility for the have-nots and of endless capacity on the part of the haves in turning privilege into their own rights. Furthermore, as far as our topic is concerned, the counterpoint of sugar and oil, this last period seems to have been characterized by what one can call a great reconciliation between these two otherwise opposing poles—the global demand for ethanol and Brazil's leading role in it have contributed to making the cultivation of sugar also part of the world of fuel. Petrobrás is more active than ever in the economics of ethanol, by distributing it, adding it to gasoline, and codetermining its price. This adds up to shaping a fairly new horizon for demands of emancipation, equality, and citizenship, which is largely related to a radical change in the self-perception of the nation—since Lula's first government in 2002 Brazil, in the eyes of the majority of the population, is not any longer a poor nation but a growing actor in the Global South now part of the BRICs, possibly the sixth-largest GNP, and is in the process of becoming a new middle-class country. The left-leaning government, through an ironic twist, rather than suggesting measures for the redistribution of the existing wealth—which would mean taking from the rich few to give to the much poorer majority—hints at the possibility of giving more to each and every Brazilian, thanks to the rising price of commodities that Brazil is bestowed with and the apparently huge reservoirs of yet to exploit deep-sea oil. This 'natural wealth' of Brazil plus the set of measures against extreme poverty and the support to the internal market and consumption have created what now the mass media call a rising new middle class—in many ways a new working class by Western European standards and ways of measuring wealth. Regardless whether this is all 'true' or also corresponds to the successful narrative of the government—and I might suggest that here we have a bit of both—one has to agree that this context shapes new sets of expectations and new opportunities for identity formation. The emphasis on (great) new sources of national wealth makes it possible for the government and the state to promise welfare for all, even at times of a sequence of crises of the global economy. As opposed to countries where wealth and identity have been presented as something to be distributed or assigned according to a zero-sum-game—that is, if one section or group wins something, another section or group will have to give in an equal amount of wealth or resources—Brazil is now proposing a new middle-class horizon for the majority of its citizens and a better life for all. This might still be far away, but minimum wages have never been so high, supplementary benefits for poor families so carefully distributed and numerous, and grants or loans for attending university so well accepted and actually used.

When one combines this with a new process of positive evaluation of popular cultures and ethnic identities through the patrimonialization of intangible culture by several sections of the state, and the favorable attitude of the Supreme Federal Court towards land claims of *índios* and maroon communities or the recent unanimous endorsement by such court of the constitutionality of affirmative action measures in higher education, we see clearly that a new horizon for emancipation and identity formation is in sight. This new and real revolution of rising expectations when combined with the perception of the new economic 'reality' has real enough outcomes on daily life, such as rising consumption of electronics and communication technologies now also in the lower classes as well as a change in attitude among young people in the lower classes towards the kinds of jobs that are in fact available for them. A growing number of people, especially young people because they are better attuned to global cultural flows and generally more interested in exploring social identities, now believe that social mobility and emancipation can proceed in huge leaps rather than gradually as they had been told. Whether this new attitude will be a motive for a new pursuit of emancipation or result in a renewed frustration is still to be seen.

NOTES

1. My title is an oblique reference to Fernando Ortiz's classic book *Cuban Counterpoint of Tobacco and Sugar* (1946), where he refers to tobacco as being the engine of a 'soft' form of production whereas sugar is the engine of a 'hard' form of production. The metaphor of the counterpoint between two 'opposing' commodities is also used, in the case of the Reconcavo region of Bahia, in Barickman (1998).
2. I thank the National Research Council (CNPq) and the Millennium Institute on Inequality based at IUPERJ, Rio, for the not exactly generous, but certainly much-needed financial help with this research project. I also thank my research assistants Washington de Jesus, Agrimaria Mattos, Evelim Sousa, and Diogenes Barbosa.
3. See http://www.ibge.gov.br/cidadesat/painel/painel.php?codmun=292920#. For a rather comprehensive book on this region, see the recent compilations edited by Caroso, Tavares, and Pereira (2011).
4. In Brazil there has been surprisingly little socio-anthropological research on sugar and, especially, oil. With few exceptions, such as José Sergio Leite Lopes (1976) who focused on one sugar mill in the state of Pernambuco, the little research that there is tends not to focus on one specific sugar mill or oil camp; largely because there are few written sources for such small-scale research, and ethnography or oral history have not been used sufficiently.
5. See, among others, Manuel Diegues Jr. This kind of popular historical account was widely read and became quite influential in the process of nation building in the period 1930–60. For a critical review of this approach and its political implications, see B. J Barickman (1996, 1998).
6. Soap, of course, is yet another global commodity with local histories and meanings.
7. See, among others, the book *Os desbravadores*, a celebratory album of fifty years of the RLAM Refinery, newspaper articles by Montero Lobato, and the recent glossy book on the history of oil in Bahia edited by Cid Teixeira (2011).
8. In many ways, I propose a theoretical approach that combines two classic attempts to identify transnational similarities within and across the different colonial styles and ecumenes of the Black Atlantic. If I could, I would bring together the insights of Charles Boxer's focus on the culture of colonization, centered on highlighting the specificity of the Portuguese seaborne empire and accounting for its sets of singularities when compared with other empires and colonial styles, with Sidney Mintz's commodity ethnography—its network, power structure, economy and culture. It would be like, finally, reconciling two contradictory tendencies:
    1. The undeniably specific traits of Portuguese colonial style—which managed to produce in a set of locations that were very much afar from one another a fairly similar culture of domination hinging upon a quaint combination of violence and intimacy.
    2. The immanent brutality and ruthlessness of global commodities—which break through the diverse colonial styles and 'culture areas'. Perhaps one can consider these two apparently contradictory tendencies as two influences whereby one mediates the other.
9. See the recent work of the sociologists Jessé Souza and José Mauricio Domingues.
10. This is an expression I borrow from the French anthropologist Michel Agier (2001), who speaks of great ethnic projects: the ones that seem to benefit

best from the forces of globalization. Not all ethnic projects benefit from the process, only those that are somehow exportable, because they are not inherently related to a specific territory, for example through a system of genealogy.

REFERENCES

Agier, Michel. 2001. "Distúrbios identitários em tempos de globalização." *Mana* 7 (2): 7–33.
Barickman, B. J. 1996. "Resistance and Decline: Slave Labour and Sugar Production in the Bahian Reconcavo, 1850–1888." *Journal of Latin American Studies* 28 (3): 588–633.
———. 1998. *A Bahian Counterpoint: Sugar, Tobacco, Cassava and Slavery in the Recôncavo 1780–1860*. Stanford, CA: Stanford Universiity Press.
Bosma, Ulbe, Juan Giusti-Croder, and G. Roger Knight, eds. 2007. *Sugarlandia Revisite: Sugar and Colonialism in Asia and the Americas, 1800–1940*. New York: Berghahn Books.
Brubaker, Rogers, and Frederick Cooper. 2000. "Beyond Identity." *Theory and Society* 29 (1): 1–47.
Caroso, Carlos, Fatima Tavares, and Claudio Pereira, eds. 2011. *Bahia de Todos os Santos: Aspectos Humanos*. Salvador: EDUFBA.
Chor Maio, Marcos. 1999. "Projeto UNESCO e agenda das ciências sociais no Brasil dos anos 40 e 50." *Revista Brasileira de Ciências Sociais* 141.
Comaroff, John. 2001. *Millennium Capitalism and the Culture of Neo-Liberalism*. Durham, NC: Duke University Press.
Coronil, Fernando. 1997. *The Magical State: Nature, Money, and Modernity in Venezuela*. Chicago: University of Chicago Press.
Costa Pinto, Luis da. 1958. *Recôncavo: Laboratório de uma experiência humana*. Rio de Janeiro: Centro Latino-Americano de Pesquisa em Ciências Sociais.
de Azevedo Brandão, Maria, org. 1998. *Recôncavo da Bahia: Sociedade e economia em transição*. Salvador: Fund. Casa de Jorge Amado.
Farias, Patrícia. 2003. "O homem *offshore*—reflexões em torno da construção do masculino e do feminino entre trabalhadores do petróleo em Campos, RJ." Paper presented at the V RAM, Florianópolis, December 3.
Fraga, Walter. 2007. *As encruzilhadas da liberdade*. Campinas: Editora da Unicamp.
Gilroy, Paul. 1993. *The Black Atlantic: Modernity and Double Consciousness*. London: Verso.
Gleason, Philip. 1983. "Identifying Identity: A Semantic History." *The Journal of American History* 69 (4): 910–31.
Handler, Richard. 1994. "Is Identity a Useful Cross-Cultural Concept?" In *Commemorations: The Politics of National Identity*, edited by John R. Gillis, 000–000. Princeton, NJ: Princeton University Press.
Harris, Marvin. 1958. *Town and Country in Brazil*. New York: Columbia University Press.
Hobsbawm, Eric. 1997. "Identity Politics and the Left." *New Left Review* 1: 217.
Hutchinson, Harry William. 1957. *Village and Plantation Life in Northeastern Brazil*. Seattle: University of Washington Press.
Karl, Terry Lynn. 1997. *The Paradox of Plenty. Oil Booms and Petro-States*. Berkeley: University of California Press.
Leite Lopes, José Sergio. 1976. *O Vapor do Diabo: O Trabalho dos Operários do Açúcar*. Rio de Janeiro: Paz e Terra.

Lomax, Allan. 1970. "The Homogeneity of African-Afro-American Musical Style." In *Afro-American Anthropology*, edited by Norman Whitten and John Szwed, 181–202. New York: The Free Press.

Margolis, Maxine. 1975. "The Ideology of Equality on a Brazilian Sugar Plantation." *Ethnology* 14 (4): 000–000.

Mattos, Hebe. 2005. *Memórias do Cativeiro: Família, trabalho e cidadania no pós-abolição*. Rio de Janeiro: Civilização Brasileira.

Mintz, Sidney. 1985. *Sweetness and Power: The Place of Sugar in Modern History*. New York: Penguin.

Pereira, Claudio & Livio Sansone eds. 2007. *O projeto Unesco no Brasil*. Textos críticos. Salvador: EDUFBA.

Sansone, Livio. 2002a. "Não trabalho, cor e identidade negra: uma comparação entre Rio e Salvador." In *Raça como retórica: a construção da diferença em perspectiva comparada*, edited by Yvone Maggie and Cláudia Rezende, 000–000. Rio de Janeiro: Record.

———. 2002b. "Remembering Slavery from Nearby: Heritage Brazilian Style." In *Facing Up to the Past: Perspectives on the Commemoration of Slavery from Africa, the Americas and Europe*, edited by Geert Oostindie, 000–000. London: Ian Randle/James Currey.

———. 2003a. *Blackness without Ethnicity*. New York: Palgrave Macmillan.

———. 2003b. "Jovens e oportunidades, as mudanças na última década e as variações por cor e classe—não se fazem mais empregadas como antigamente." In *Desigualdades sociais: o estado da nação*, edited by Carlos Hasenbalg and Nelson do Valle Silva, 000–000. Rio de Janeiro: Topbooks.

———. 2012. "Turner, Franklin and Herskovits in the Gantois House of Candomble: The Transnational Origin of Afro-Brazilian Studies." *Black Scholar* 41 (1): 48–63.

Schwarz, Stuart. 1995. *Segredos internos*. São Paulo: Companhia das Letras.

Tilly, Charles. 1998. *Durable Inequality*. Berkeley: University of California Press.

Vale de Almeida, Miguel. 1999. "Poderes, Produtos, Paixões: O Movimento Afro-Cultural numa Cidade Baiana." *Etnográfica* 3 (1): 131–56.

Viveiros, Mara. 2002. *De quebradores y cumplidores: sobre hombres, mascuklinidades y relciones de genero en Colombia*. Bogotá: Ed. Universidad Nacional De Colombia, Centro De Estudios Sociales.

Wagley, Charles. 1960. "Plantation America: A Culture Sphere." In *Caribbean Studies: A Symposium*, edited by Vera Rubin, 3–13. Amsterdam: Publisher.

———, ed. 1963. *Race and Class in Rural Brazil*. Paris: UNESCO.

Wagley, Charles, and Cecilia Roxo. 1970. "Serendipity in Bahia." *Universitas* 6 & 7 (May/December).

11 Why (Post)Colonialism and (De)Coloniality Are Not Enough

A Post-Imperialist Perspective[1]

Gustavo Lins Ribeiro

The need to examine knowledge production in relation to location and subject position is a consolidated trend in several theoretical approaches. In fact, this is a well-known postulate within the sociology of knowledge and its acceptance does not necessarily imply a critical standpoint. Perhaps the novelty of the past decades has been the great visibility and usage of frameworks that put emphasis on (a) the inquiry of the relationships between knowledge production and politics; (b) the inquiry of the relationships among locations, subject positions, and power; and (c) how taxonomies are functional and inherent to the exercise of domination. This is a trend well epitomized by the classic work of Edward Said on Orientalism (Said 1979) and by the many interpretations and debates that came to be known under the umbrella label of postcolonialism and others such as the geopolitics of knowledge (Mignolo 2001).

Indeed, the intellectual and political sensitivity to the complex relationships between difference and power is now widely diffused. It is a consequence of political struggles and macro global processes that in larger or lesser degree made the politics of identity, multicultural policies, and expressions such as the 'West and the Rest' and the 'Global South' part of everyday life within and without academia. These taxonomic devices substituted for older ones, like the 'Third World', that kept an obvious link to the Cold War juncture. In the current juncture, the particular/universal tension was submitted to a new round of criticism and, almost everywhere, knowledge producers make new claims to visibility and validity, new claims to the empowerment of a variety of world visions. Such struggles happen in a milieu structured by the hegemony of Western knowledge. The quandaries are particularly intense when the human and social sciences are involved because they are sensitive to context and meaning, to ideology and utopia, to the definition of destiny and the good life. The last and hottest frontier of these clashes is the production of knowledge by indigenous populations. They have become subjects of their own epistemological and philosophical struggles influencing academic and political life in different degrees and places—see, for instance, Ecuador (Walsh 2002, 2007), Bolivia (García Linera 2008), and New Zealand (Smith 1999).

Will all agents of knowledge production—especially in those loci where decolonization is an issue—struggle to prove the equality or superiority of their local knowledge vis-à-vis the Western and other less powerful loci of knowledge production? Are we going to end up with a Babel of knowledge claims? This is not my vision. I believe that we will see new forms of conviviality among epistemologies, paradigms, and approaches. As an anthropologist I cannot believe in total incommensurability among mind-sets and interpretations, a position that does not amount to a naïve acceptance of a transcendental universality. I am aware that most claims to universality are based on power effects. In a globalized world, the problem is the imperial pretension to hegemony, the imposition of viewpoints that are disseminated through painless structures of prestige diffusion from global or national hegemonic centers. However, one thing is a claim to universality based on power relations; another thing is a claim to universality based on empathy, sympathy, sharing, and the art of argumentation and convincing. The more different subject positions proliferate and experiences of horizontal exchanges within the world system of knowledge production exist the better for all of us.[2] Anchored on more diverse grounds, the resulting cross-fertilization will be more complex and capable of surpassing the current monotony of the Anglo-American academic hegemony.

In this regard, the free-software movement may provide a source of inspiration with its global cooperation and articulation of an enormous amount of global fragmented agencies and spaces. Open-source publishing allows us to speculate about the possibility of a wiki-anthropology, for instance, one that would outdo the traditional journals with their referee system, which, in the core of the world system of anthropological production, more often than not replicate the styles and agendas of the Anglo-American academic milieu (Mathews 2010). Global online publications with free access already exist and may potentially change the hierarchy of journals' visibility and prestige. The possibility of writing with a myriad of other known or anonymous cyber-colleagues may also point out to the emergence of postauthorial academic texts. Are we ready to make global wiki experiments in academic writing and theoretical production? Are we ready to go beyond the notion of authorship in academia, one of the basis of inequality reproduction in a world full of individualism and individual power seekers? I don't know. Perhaps my generation is not. Perhaps younger scholars, natives of digital culture completely immersed in cyberspace, are.

Incommensurability is a key word here. Am I incapable of understanding what an Indian intellectual writes in Delhi or the intertextuality of the formulations of the Yanomami leader Davi Kopenawa (Albert 1995)? And what of the wide influence and appropriation of Western thought produced by, say, Marx, Durkheim, Weber, Freud, and many others almost everywhere? Are they only a sad portrait of intellectual misery and sub-alternity within the world system of knowledge production? A positive answer to these questions would mean the insularity of theory. Theory would be imprisoned in a myriad of places, it would not travel, and we know it is in the nature of theory to travel. More than twenty years ago, James Clifford wrote:

> Theory: returned to its etymological roots, with a late twentieth-century difference. The Greek term *theorein*: a practice of travel and observation, a man sent by the polis to another city to witness a reli-gious ceremony. 'Theory' is a product of displacement, comparison, a certain distance. To theorize, one leaves home. But like any act of travel, theory begins and ends somewhere. In the case of the Greek theorist the beginning and ending were one, the home polis. This is not so simply true of traveling theorists in the late twentieth century. (Clifford 1989)

Were theories to be reduced to a series of local autarchies, no one would ever learn from other people, especially from the most distant ones. Instead, I believe that in a globalized world we are all forced to look for working commensurabilities that open communication channels among different semantic universes. What we need is more "traveling theories" and not only those with Western accents (Said 1992). This is why I consider the role of anthropology to be even more central today than in the past. What is unac-ceptable, let me repeat to emphasize, is an imperial pretension to universal-ism whether it comes from the Global North or from the Global South.

I also find troublesome the role of 'global reception machine' played by North American academia today. This is why, in the past, I have argued that in Latin America we cannot uncritically accept the dissemination of theories, such as postcolonialism, that arrive in the region after being more or less indigenized in the US:

> If colonial discourse analysis and post-colonial theory are "critiques of the process of production of knowledge about the Other" (Williams and Chrisman, 1994: 8), it would be at least ironic that post-colonial-ism—with its trajectory marked by its growth and proliferation in Eng-lish-speaking academia—colonizes—if you excuse the wordplay—the empty space left by the absence of Latin American cosmopolitics and becomes a discourse to produce knowledge about the Latin American Other. In Latin America post-colonialism would be equal to what it condemns, a foreign discourse on the Other that arrives through the hands of a metropolitan power. Post-colonialists would be, unwittingly, doing what they criticized. Obviously, post-colonialism's dissemina-tion cannot be reduced to the force of the Anglo-American hegemony behind it. Similar to other critical cosmopolitics, post-colonialism has contributions to make in the analysis of social, cultural and political

realities anywhere, especially when power asymmetries are at stake. The issue is not to deny post-colonialism but to assert the production of critical narratives in tune with Latin American subject positions, in a heteroglossic dialogue with cosmopolitics from other glocalities. (Ribeiro 2008: 38)

For me, the notion of cosmopolitics is central to understand the current production of theories and disciplines that pretend to have global reach (Ribeiro 2006; Cheah and Robbins 1998). It is based, on the one hand, on positive evocations historically associated with the notion of cosmopolitism and, on the other hand, on analysis in which power asymmetries are of fundamental importance. Cosmopolitics comprises discourses and modes of doing politics that are concerned with their global reach and impact. Several cosmopolitics are counterhegemonic discourses anchored on particular situations. This is the case with postcolonial critique, decoloniality of power, Zapatismo, subaltern studies, and *interculturalidad,* a perspective that is being more clearly elaborated in the Andes, especially in Ecuador (Walsh, Schiwy and Castro-Gómez 2002; García Canclini 2004; Rappaport 2005; Bartolomé 2006). Because there are several progressive cosmopolitics, articulation becomes a key word. Indeed, the effectiveness of cosmopolitical initiatives on the transnational level relies on networking. There is not a singular cosmopolitics capable of dealing with the entire complexity of the global counterhegemonic struggle and with the existence and proliferation of critical subjects in fragmented global spaces. Supporters of different counterhegemonic cosmopolitics need to identify their mutual equivalences to be able to articulate themselves in networks and political actions. Effective nonimperialist cosmopolitics that inform transnational political activists and progressive forms of global awareness also require a complex articulation of multilocated and plural struggles and subjects.

The notion of cosmopolitics greatly coincides with one kind of particularism I call "cosmopolitan particularism," i.e., "discourses that intrinsically address global issues and pretend to be taken into consideration, if not incorporated, by other people" (Ribeiro 2007: 14). "Cosmopolitics" allows me to explore cosmopolitan particularisms as a form of global political discourse and to go beyond the particular/universal tension that, in one way or another, is a grid framing this discussion. There are several cosmopolitics that may complement each other in a complex heteroglossic conversation of equivalencies. It is within such a framework that I envisage the relations among postcolonialism, decoloniality of power, and the approach I call postimperialism (Ribeiro 2003, 2008).

Postimperialism is the label I use to define the current juncture in which nation-states have to deal with transnationalism and with the effects of flexible capitalism. This is a kind of transnationalism marked by an intense time-space compression (Harvey 1989), i.e., by a technological command of space and time that distances itself more and more from the political and administrative forms of exerting power associated with modern imperialism and from the colony in its strict sense of occupation of a foreign land. Undoubt-edly, postimperialism exists with other forms of organizing economic and political life and constructing cosmopolitics. After 9/11, imperialism has resurged in Afghanistan and Iraq, a fact that shows, once again, that history does not move in a straight line and that the conservative military-industrial complex has known very well how to maintain its power and take advantage of certain political opportunities in the US. However, in Latin American nation-states, the political independence of which started in the first decades of the nineteenth century, postimperialism predominates over other dynam-ics. It informs the contents of political, economic, and cultural contempo-raneity as well as imposes certain interpretative and research needs. I want to advance the idea that postimperialism is the Latin American side of the coin on which postcolonialism is found. It should be clear that I use the term ironically. Furthermore, as a cosmopolitics, postimperialism mixes utopian horizons (a moment beyond imperialism in which, nonetheless, imperialism remains an issue) and descriptions of specific characteristics of our times. It thus combines programmatic and sociological visions.[3]

Currently, a number of some of the most interesting Latin American intellectuals—a few of them long established in North American universi-ties—see themselves as members of a "school of thought" that takes the dis-cussion about "de-coloniality of power" as an organizing and congregating axis (Grosfoguel 2009).[4] The complicated and difficult relationships between Latin American studies, postcolonialism, and de-coloniality have been out-lined in a book edited, in the US, by Mabel Moraña, Enrique Dussel, and Carlos A. Jáuregui (2008). It is not my intention to go over the intricacies and complexities of this debate here. Rather, my goal is to explore the idea that the locus of enunciation on academic subjects is geopolitically marked. In this connection, it is impossible not to recognize a strong Andean (and secondarily Mexican) accent in the de-coloniality of power cosmopolitics.

I want to make explicit that my own positionality, like many others, reflects different itineraries and engagements. I will highlight but a few, those more relevant for my arguments here: (1) the experience of grow-ing up in the high-modernist federal city of Brasilia after its inauguration in 1960; (2) the resistance against the Brazilian military dictatorship of 1964–85; (3) the writing of a history of the construction of Brasilia from the workers' point of view to criticize the nationalist ideologies covering the construction of the city (Ribeiro 2008a); and (4) my graduate education in a Brazilian and a North American university.

THE MANY LIVES OF COLONIALISM AND ITS RESURRECTIONS

The historical, geographical, economic, cultural, political, social, and racial diversity of the colonial experience is, sometimes, underestimated by an overall critique of one of the most powerful attributes of the human species:

the propensity to colonize the entire earth. Human beings could well be called the 'colonizer animal'. If we stretch the amplitude of the colonial drive that much, a substantial part of the history of Homo sapiens could be seen as the history of colonization and of its ideological, institutional, and political legitimations. In this time frame, one that perhaps could be called an archaeological time frame, postcolonial situations would include an enormous number of scenarios. Such a perspective would allow me to go as far as to call, for instance, London a postcolonial city because one day it was a Roman town called Londinium.

I will not concede to the temptation of this overarching vision because it would certainly transform colonialism into a sort of hyper-historicism. Furthermore, the duration of the postcolonial condition is an area of intense debate, one that is particularly interesting for Latin Americans because, historically, postcolonialism started there in the early nineteenth century. Here I want to tackle with the consideration of colonialism as a powerful, all-encompassing historical force that shapes to these days major characteristics of nation-states that are former colonies. Colonialism and its effects have been in the center of critical analysis well before postcolonialism and de-coloniality became focuses of attention. Neocolonialism and internal colonialism provided important theoretical frameworks to analyze the existing inequalities within the world system and within nation-states.

The stress on colonialism, neocolonialism, internal colonialism, postcolonialism, and the coloniality of power is welcome. No one doubts the power of structuration of colonialism. However, I'd like to explore the idea that we cannot think of the 'structural power' of colonialism as a lasting force that *always* overruns others, especially those that are unleashed by what might be called 'the nationality of power' (for the notion of structural power, see Wolf 1999). For me, postcolonialism and the coloniality of power coexist in different forms and intensities, in different national scenarios, with the indigeneity of power, with the nationality of power, as well as with the globality of power. On the one hand, however strong transnational forces may be, we cannot diffuse the power of nation-states in global entities such as the world system nor reduce them to mechanic responses to supranational dynamics. On the other hand, colonialism cannot become an interpretive panacea nor the latest example of historical determinism. Lastly, the characteristics of previously existing systems of indigenous power cannot be disregarded. It is one thing to colonize former empires, like the Aztec and Inca ones; it is another to colonize an area of hunters and gatherers (like the one that is Brazil today).

The fact that peripheral countries are the privileged scenario for postcolonial and de-colonial interpretations becomes a problem when we realize that the most powerful nation-state of current times, the US, is a former British colony. If the explanation for this exception is that there are different colonial experiences that may result in different postcolonial and de-colonial experiences, then subalternity within the world system is not a necessary result of the colonial experience nor an intrinsic quality of post-coloniality and de-coloniality. What I am saying is that an overemphasis on colonialism and on coloniality can curiously (re)generate precisely what needs to be criticized and surpassed: an explanation that accepts subalternity as a destiny of former colonies.

My argument calls for a sharper consideration of the 'causal hierarchies' among colonialism and other historical processes in diverse concrete scenarios (for the notion of causal hierarchies, see Heyman and Campbell 2009). I am implying that by transforming colonialism and not capitalism into the primordial focus of analysis, we underestimate the current importance of nation-states and their elites, as well as deviate from understanding the particular characteristics of the power relations of the current relationships between nation-states and the world system. In some places, these relationships are 200 years old or more, if we include in our list the US, the first politically independent modern republic. Isn't this a sufficient amount of time to create particular interests and dynamics that are central to the construction of any future scenario?

If one of the aims of critical theory is to overcome an unjust past and contribute to the construction of a different future, then utopias are a most important object of desire in the progressive intellectual scene. Although I am favorable to ideological struggles—without them it wouldn't be possible to denaturalize the naturalized present—I want to advocate for more utopian struggles in a juncture where there is a dearth of future scenarios strong enough to galvanize the imagination of a great amount of political actors. This is one of the reasons why I offered the notion of postimperialism. Living in a world region that has a long-standing experience with imperialism—in its soft and hard expressions—the imagining of life after imperialism can prove to be an exercise in creativity and audacity—qualities many times denied to the 'subalterns'.

In order to clarify how different colonial enterprises may be experienced in history and how they may shape different senses of the future, I will recur to a reading of the Brazilian postcolonial and national history. My arguments suppose that colonialist 'structural power' coexists not only with world system forces but also with the rise of postcolonial projects that may congeal and prompt the nationality of power.

What follows is not an exercise in 'methodological nationalism'. Quite the contrary, it is a cosmopolitical exercise. I can argue the way I do only because I constantly relate my own cosmopolitan particularism to other cosmopolitan particularisms produced elsewhere in the world.

THE POSTCOLONIAL LIFE OF A TROPICAL EMPIRE

The Brazilian colonial experience differs from the experience of other countries elsewhere in Latin America. Brazil was the only country colonized by the Portuguese in a large area of lowland South America inhabited by indigenous peoples that were not organized, as in the Andes and in Mexico, under powerful native empires. The early nineteenth century provides a most interesting period to further develop my arguments. It is the time when most Latin American countries were starting their political independence and a true postcolonial period. Although formal political independence in Brazil started only in 1822, more than ten years later than several of its neighbors, I want to submit the idea that the 'postcolonial' Brazilian period started in 1808 under the impact of European power struggles meant to define hegemony within the world system. The invasion of Portugal by Napoleonic troops forced the court of Portugal to flee to Brazil. In November 1807, some 15,000 people crossed the Atlantic under the protection of the British Navy in what is perhaps the largest forced migration of political elites in world history.

What is even more special about this move is that the capital of the Portuguese empire was transferred to Rio de Janeiro in 1808 where the king of Portugal, Don João VI, and the court went to live and stayed for thirteen years until 1821. I am not aware of any other example in the history of imperialism/colonialism (certainly not in the Americas and in Western imperialism) of a colony that is suddenly transformed into the seat of the empire. The colonial status of Brazil was terminated in such an unusual way that calling the 1808–21 period a postcolonial period may not be very accurate. First, to become the center of the empire is not a postcolonial condition; second, after the return, in April 1821, of Don João VI to Lisbon, some of the typical contradictions between imperial centers and colonial peripheries arose again. They lasted for a short period of less than eighteen months and forced Prince Don Pedro, who was the son left behind by Don João to take care of the affairs of the crown in Brazil, to declare the country's political independence in September 1822 and become the emperor Pedro I.

The Brazilian case is a complex mélange of continuity and discontinuity, a major *problématique* when the prefix post is at stake as in postcolonialism. But I will consider 1808 as the starting year of postcolonialism in Brazil anyhow for it was the year when Brazilians began to experience administrative autonomy and, most importantly, the ruling elite started to consider leaving behind the territorial pattern—a powerful factor in colonial dominance—structured by the colonial regime, a pattern that had created highly secluded and barely integrated regional systems turned and adapted to Europe's needs. Brazilians, it would be said, are like crabs—they cling to the Atlantic shoreline. Indeed, already in 1808 the king of Portugal and his advisers considered the possibility of moving the capital to the hinterlands. This idea was first linked to what was considered Rio's urban inadequacies to be the capital of the Portuguese Empire: its climate and its lack of adequate infrastructure, for instance. A cosmetic Europeanization of Rio was promoted then.

But the critique of Rio was also a means to criticize colonial life and would soon converge to elaborate geopolitical debates (Vidal 2009). In different ways and corresponding to different political and economic interests, most discourses of the time pointed out the need to structure a new nation, an 'empire', from within the continental territory controlled by Portugal. Hipólito José da Costa, for instance, in 1813, wrote that:

> If the courtiers that went . . . from Lisbon [to Rio] had enough patriotism and acknowledgement for the country that received them, they would make a generous sacrifice of the comfort and luxury they may enjoy in Rio de Janeiro and would establish themselves in the hinterland and central areas close to the headwaters of the big rivers; they would build a new city there and would start to open new roads headed towards the maritime harbors and would remove the natural obstacles that the different navigable rivers have, and would thus lay the foundations of the most extended, connected, well defended and powerful Empire that can possibly exist on the earth of the Globe, according to the current state of the nations that people it. (Vidal 2009: 41)

French historian Laurent Vidal (2009) states that the critique of the lack of regional integration was generalized in early nineteenth-century Brazil but was particularly strong among landowners who wished to optimize their use of agricultural and other economic resources with a view to expanding their power to the whole country. From 1821 to 1824, when the first constitution of independent Brazil was promulgated, the discussion about the capital's transfer revolved around how to "symbolically mark the passage from the Portuguese Empire to the Brazilian Empire. The task was, in short, to 'decapitalize' a colonial and maritime Empire to 'recapitalize' an independent and continental Empire" (Vidal 2009: 51). The construction of the new city was thought as a rupture with the metropolis, a geopolitical decision that should be based on visions of the future of the nation meant to foster and consolidate national unity and progress (Vidal 2009: 52 ff.).

The use of the term 'empire' is not by chance. After the declaration of Brazil's political independence, in 1822, the son of the king of Portugal, Don Pedro, as mentioned before, became the first Brazilian emperor and the country an 'empire'. As the interests of pro-Portugal new Brazilian elites became more influential during the discussion of the first Constitution of 1824, the idea of leaving Rio fell into the background. Interestingly enough, Brazil was ruled by the same royal house as in Portugal, the Bragança, for sixty-seven years, until the declaration of the Republic in 1889. It might be said that this peculiar transition from colonial to postcolonial times contributed to the maintenance of Brazil's territorial unity in a stark contrast to the results of the Bolivarian and San Martinian projects of a unified postcolonial Spanish speaking South America. It also might be said that such a peculiarity meant a rather different postcolonial experience vis-à-vis the other Latin American countries that moved from the colonial condition, i.e., from the subordination to a royal European power, to the republican form of government. Brazil was the only country in the Americas that became a monarchy after its independence. Along the Brazilian monarchy period the idea of moving the capital to the hinterlands did not disappear. Quite the contrary, it was so much alive that it became a formal provision of the first Republican Constitution of 1891. The need to construct a new capital in the center of the country was now a goal of postimperial Brazil.

During the first half of the twentieth century, a few initiatives were taken and prepared the scenario for the moving of the capital. Even though the transference had been formally considered by the ruling elites since the early nineteenth century, only in 1960, more than 150 years later, the capital was finally transferred to Brasilia. The continuity of the capital in Rio was an expression of the continuity of colonialism's structuring power.

THE NATIONALITY OF POWER OF A POSTIMPERIAL CITY

The transference of the Brazilian capital allows me to think in hierarchical causal terms. Whereas the hegemony of the postcolonial structuration ends with Brasília, the prominence of the nationality of power starts at the same moment. In 1960, as in 1808, the main goal of the moving of the capital was to integrate the country from within. Along with the construction of the new city, new roads were open to link all regions by land. Now the colonial regional systems and the expanding capitalist agricultural frontiers would have to coexist with other internal dynamics related to the creation of an integrated national territory, actually an integrated capitalist national economy/market, from within the hinterland of the country. Norbert Elias has pointed out the importance of territorial integration for nation building: "Societies become nations when the functional interdependency among their regions and social strata, as well as among their hierarchical levels of authority, becomes sufficiently large and reciprocal so that none of these groups may completely disregard what the others think, feel or wish" (Elias 2006: 163).

Predictably, nationalism was the main ideological force behind the transfer of the capital and the source of legitimating discourses. Juscelino Kubitschek, the president (1956–60) that led the construction of Brasilia, is, to these days, the most popular president of Brazil. But inasmuch as the role of national political leaders was central to the process, the deepest movement underneath the capital's transfer was the expansion of the agricultural frontiers westward, to the huge savannah area, an ecosystem that was almost entirely destroyed with the new expansionists moves the construction of Brasília generated. The savannah area became a stepping-stone for the colonization of the Amazon region and is currently a major exporter of soybeans and beef.

Brasília is the only capital city in the Americas that is not built over or adjacent to a former colonial settlement. What did Brazilian national elites wanted to say when they built the new capital? They certainly did not want to say the country was the hyper West, as North Americans seemed to say with the oversized neoclassical architecture of some of the major federal buildings in Washington. Brazilians wanted to affirm their difference, that they were modern and in charge of their own history. The fact that Rio— the metonym of the tropicalist-colonialist image of Brazil, stereotyped as the land of natural exuberance and sensual happy-go-lucky natives (Ribeiro 2004)—was left behind in favor of a city that represented a sum of creativity, ingenuity, and toil was contradictory with the prevailing Orientalist view of Brazil (including within the academic milieu). The force of the nationalist claim materialized in a large-scale project is the strongest index of the change in the relations between postcolonial and national forces, between the coloniality and the nationality of power. The construction of a futurist and utopian city intended to send the message that the future was happening in the hinterlands of Brazil. This is a particularly effective trope given the importance of the future and of scale in the nationalist imaginary of Brazil, the giant "eternally laid in a splendid cradle," a well-known phrase of the national anthem.

FINAL REMARKS

In this section, I will make a few general concluding remarks and will draw conclusions that are specific to the Brazilian scenario but that also relate to the need to further develop postimperialist perspectives. In spite of the power of structuration of colonialism, it cannot be seen as an overall force determining all current sociological, economic, political, and cultural scenarios in previously colonized nation-states. The duration of the postcolonial period and the prominence of the coloniality of power vary in different historical settings. The definition of such moments needs to be found on a case-by-case basis. I would argue that in Bolivia, for instance, the moment of shift from the prominence of the coloniality of power to the beginning of the construction of the nationality of power happened only with the election of Evo Morales as president in 2006. This leads me to think that the close relation between the formulation of the theory on the coloniality of power and the political life of Andean countries such as Bolivia, Peru, Ecuador, and Colombia is an index of the relative strength of the power of structuration of colonialism in these countries. In view of the variability of global historical experiences the 'nationality of power' cannot be subsumed under colonial frameworks of analysis nor under globalized ones, it is a specific object of inquiry. Therefore a more complete framework of analysis includes causal hierarchies that are sensitive to the different geographies and histories of colonialism and of nation building the power of structuration of which varies over time according to the outcome of different historical conflicts in different nation-states. In sum, former colonies are differently subject, today, to the diverse powers of structuration stemming from the indigeneity of power, the coloniality of power, the nationality of power (that includes the histories, specificities, and contradictions of the local and regional levels), and the globality of power. All of these need to be understood in the framework of an ever-expanding capitalist political economy with its dynamics and contradictions.

Throughout the postcolonial and national history of Brazil, a strong ideology of the ruling elites developed according to which the country is destined to become a world power. The construction and consolidation of Brasília as the country's new capital were a crucial step in the development of the Brazilian nationality of power, they reassured the 'great destiny of Brazil' to nationalist ideologues and reinforced the discursive matrix of a powerful future. In the current moment of the world system, especially after the 2008–9 crisis when the BRICs became the most publicized examples of fast response to the crisis, the sense that the 'sleeping giant' is about to wake up increased within Brazilian political and economic elites. It is already possible to see that Brasília will become in the near future the capital city of a major global player with part of its elite with (sub) imperialist pretensions.[5] The role of critical thought in Brazil in this regard is to make a preemptive move in order to go beyond such pretensions and favor the rise not only of a postimperialist capital city but also of a postimperialist country. By this I mean a kind of cosmopolitics that imagines a world system without imperialisms and fosters national formulations and actions in international arenas that stress and truly promote cooperation and peace at the same time that it criticizes inequality and war. To do that there is a need to dedicate more time to a postimperialist imagination, critique and program; to dedicate, in sum, more time to utopian struggles than to ideological ones. Postimperialism would thus be a cosmopolitics capable of pointing out to new moments of the world system and its unfolding.

NOTES

1. This is a slightly different version of a text published in *Postcolonial Studies* 14, no. 3, 285–97, 2011. I thank the editors of *Postcolonial Studies* for allowing its publication in this volume.
2. The notion of a 'world system of knowledge production' is inspired by Takami Kuwayama (2004). It is an obvious metaphor of the unequal capacities of production and dissemination of knowledge on a global scale.
3. More on postimperialism in Ribeiro (2003, 2008). Postimperialism is an ironic expression, not in the sense that I mean the opposite that it indicates but in the sense that I want to provoke a malaise in the reader to make him or her think in another way, to open a different universe of questioning and imagining.

204   Gustavo Lins Ribeiro

4. Two crucial texts on the coloniality of power are Quijano (1993) and Dussel (1993).
5. The *Correio Braziliense*, Brasilia's most important daily paper, showed on its front page in its edition of July 17, 2011, the following headline: "The Fall of the American Empire . . . and the Rise of the Brazilian Empire."

REFERENCES

Albert, Bruce. 1995. "O Ouro Canibal e a Queda do Céu: uma crítica xamânica da economica política da natureza". Série Antropologia no. 174. Universidade de Brasília.
Bartolomé, Miguel Alberto. 2006. *Procesos Interculturales: Antropología Política del Pluralismo Cultural en América Latina*. México: Siglo XXI Editores.
Cheah, Pheng, and Bruce Robbins, eds. 1998. *Cosmopolitics: Thinking and Feeling Beyond the Nation*. Minneapolis: University of Minnesota Press.
Clifford, James. 1989. "Notes on travel and theory". *Inscriptions* 5: 177-185. http://culturalstudies.ucsc.edu/PUBS/Inscriptions/vol_5/clifford.html
Dussel, Enrique. 1993. "Europa, Modernidad y Eurocentrismo." In *La Colonialidad del Saber: Eurocentrismo y Ciencias Sociales: Perspectivas Latinoamericanas*, edited by Edgardo Lander, 41–53. Buenos Aires: Clacso.
Elias, Norbert. 2006. "Processos de formação de Estados e construção de nações". In *Escritos e Ensaios*, edited by Federico Neiburg and Leopoldo Waizbort, 153-65. Rio de Janeiro: Jorge Zahar Editor.
García Canclini, Néstor. 2004. *Diferentes, Desiguales y Desconectados: Mapas de la interculturalidad*. Barcelona: Gedisa.
García Linera, Álvaro. 2008. *La potencia plebeya. Acción colectiva e identidades indígenas, obreras y populares en Bolivia*. Buenos Aires: Clacso/Prometeo Libros.
Grosfoguel, Ramón. 2009. "Izquierdas e Izquierdas Otras: entre el proyecto de la izquierda eurocéntrica y el proyecto transmoderno de la nuevas izquierdas descoloniales". *Tabula Rasa* 11: 9-32.
Harvey, David. 1989. *The Condition of Post-Modernity*. Oxford: Basil Blackwell.
Heyman, Josiah and Howard Campbell. 2009. "The anthropology of global flows: A critical reading of Appadurai's `Disjuncture and Difference in the Global Cultural Economy'". *Anthropological Theory* 9: 131-148.
Kuwayama, Takami. 2004. *Native Anthropology: the Japanese Challenge to Western Academic Hegemony*. Melbourne: Trans Pacific Press.
Mathews, Gordon. 2010. "On the Referee System as a Barrier to Global Anthropology." *The Asia Pacific Journal of Anthropology* 11 (1): 52–63.
Mignolo, Walter. 2001. "Introducción." In *Capitalismo y Geopolítica del Conocimiento. El Eurocentrismo y la Filosofia de la Liberación en el debate intelectual contemporáneo*, edited by Walter Mignolo, 9-53. Buenos Aires: Ediciones del Signo.
Moraña, Mabel, Enrique Dussel, and Carlos A. Jáuregui, eds. 2008. *Coloniality at large. Latin America and the postcolonial debate*. Durham: Duke University Press.
Quijano, Aníbal. 1993. "Colonialidad del Poder, Eurocentrismo y América Latina." In *La Colonialidad del Saber: Eurocentrismo y Ciencias Sociales: Perspectivas Latinoamericanas*, edited by Edgardo Lander, 201–46. Buenos Aires: Clacso.
Rappaport, Joanne. 2005. *Intercultural Utopias*. Durham, NC: Duke University Press.

Ribeiro, Gustavo Lins. 2003. *Postimperialismo. Cultura y Política en el Mundo Contemporáneo.* Barcelona, Gedisa Editorial.

———. 2004. "Tropicalismo y Europeísmo. Modos de Representar al Brasil y a la Argentina". In *La Antropología Brasileña Contemporánea. Contribuciones para un diálogo latino americano,* edited by Alejandro Grimson, Gustavo Lins Ribeiro and Pablo Semán, 165-195. Buenos Aires: Prometeo.

———. 2006. "World Anthropologies: Cosmopolitics for a New Global Scenario in Anthropology". *Critique of Anthropology* 26 (4): 363-386.

———. 2007. "Cultural diversity as a global discourse". Série Antropologia no. 412, Universidade de Brasília.

———. 2008. "Post-imperialism. A Latin American cosmopolitics". In *Brazil and the Americas. Convergences and Perspectives,* edited by Peter Birle, Sérgio Costa, Horst Nitschack, 31-50. Madrid/Frankfurt: Iberoamericana/Vervuert.

———. 2008a. *O Capital da Esperança. A experiência dos trabalhadores na construção de Brasília.* Brasília: Editora da Universidade de Brasília.

Said, Edward W. 1979. *Orientalism.* New York: Vintage Books.

———. 1992. *The World, the Text, and the Critic.* Massachusetts: Harvard University Press.

Smith, Linda Tuhiwai. 1999. *Decolonizing Methodologies. Research and indigenous peoples.* London & New York/Dunedin: Zed books/University of Otago Press.

Vidal, Laurent. 2009. *De Nova Lisboa a Brasília. A invenção de uma capital (séculos XIX-XX).* Brasília: Editora da Universidade de Brasília.

Walsh, Catherine. 2002. "La (re)articulación de subjetividades políticas y diferencia colonial en Ecuador: reflexiones sobre el capitalismo y las geopolíticas del conocimiento". In *Indisciplinar las ciencias sociales. Geopolíticas del conocimiento y colonialidade del poder. Perspectivas desde lo andino,* edited by Catherine Walsh, Freya Schiwy and Santiago Castro-Gómez, 175-214. Quito: Universidad Andina Simon Bolívar/Abya-Yala.

———. 2007. "Interculturalidad y colonialidad del poder. Un pensamiento y posicionamento 'otro' desde la diferencia colonial". In *El Giro decolonial. Reflexiones para una diversidad epistémica más allá del capitalismo global,* edited by Santiago Castro-Gómez and Ramón Grosfoguel, 47-62. Bogotá: Siglo del Hombre Editores/ Universidad Central/Pontifícia Universidad Javeriana.

Walsh, Catherine, Freya Schiwy, and Santiago Castro-Gómez, eds. 2002. *Indisciplinar las ciencias sociales: Geopolíticas del conocimiento y colonialidad del poder: Perspectivas desde lo andino.* Quito: Universidad Andina Simon Bolívar/ Abya-Yala.

Williams, Patrick and Laura Chrisman. 1994. "Colonial Discourse and Post-Colonial Theory: an introduction". In *Colonial Discourse and Post-Colonial Theory,* edited by Patrick Williams and Laura Chrisman, 1-20. New York: Columbia University Press.

Wolf, Eric. 1999. *Envisioning Power. Ideologies of dominance and crisis.* Berkeley: University of California Press.

Contributors

Breno Bringel is a professor of Sociology at the Institute of Social and Political Studies, State University of Rio de Janeiro (IESP-UERJ). He holds a PhD at the Faculty of Political Science and Sociology, University Complutense of Madrid, where he also teaches. He is a member of the board of the International Sociological Association Research Committee on Social Classes and Social Movements (RC-47), editor of its *Newsletter on Social Movements*, and author of several works on social movements, contemporary internationalism, and the spatialities of transnational activism. His latest book is *Os movimentos sociais na era global* (with Maria da Glória Gohn, Rio de Janeiro, Vozes, 2013).

Adalberto Cardoso, PhD in sociology (USP, 1995), is a professor at the Institute of Social and Political Studies, State University of Rio de Janeiro (IESP-UERJ). His research interests include sociology of work (class formation, labor market dynamics, work trajectories, labor movements), social inequalities, youth, urban sociology, and social theory. He has authored nine books, including *A construção da sociedade do trabalho no Brasil* ([The construction of the work society in Brazil], Rio de Janeiro, FGV, 2010); and *As Normas e os Fatos: Desenho e Desempenho das Normas de Regulação do Mercado de Trabalho no Brasil* (] The norms and the facts: design and effectiveness of labor market regulations in Brazil], Rio de Janeiro, FGV, 2007).

Sonia Fleury holds a PhD in political science, an MA in sociology, and a BA psychology. She is a full professor at the Brazilian School of Public Administration and Business, Getulio Vargas Foundation. She has written and edited 23 books and 186 articles and chapters, published in Portuguese, Spanish, English, and French. She is an activist of the social movement that pushed for sanitary reform and social rights in Brazil. She held leading positions in major civil society organizations and worked as consultant for social protection policies during the work of the National Constituency Assembly. She was invited by President Lula to be a member of the Social and Economic Development Council.

Carolina Matos is a part-time lecturer at the Government Department at Essex University. A former fellow in political communications at the LSE, Matos obtained her PhD in media and communications at Goldsmiths College and has taught and researched in the UK in political communications, Brazilian media, and politics at the University of East London, St. Mary's College, Goldsmiths, and LSE. With twenty years of professional experience as both a journalist and academic, Matos has many articles in journals and has worked as a full-time journalist in Brazil for Reuters, Unesco, Folha de Sao Paulo, Tribuna da Imprensa, and Globo.com. Matos is the author of *Journalism and Political Democracy in Brazil* (Lexington Books, 2008) and *Media and Politics in Latin America: Globalization, Democracy and Identity* (I. B. Tauris, 2012).

Marcelo Medeiros does research on social inequality. He is currently a professor at the University of Brasília and senior researcher at the Institute for Applied Economic Research (IPEA). He has publications in the areas of gender, poverty, social mobility and inequality, health, education, time use, disability, and social protection. He received the Fred L. Soper Award for Excellence in Public Health Literature 2012 from the World Health Organization, a medal of the Brazilian Senate for development studies in 2000, the J. A. Rodrigues prize for the best PhD Thesis 2003 from the Brazilian Social Sciences Association, and two other academic prizes.

Erica Mesker received her MA in global and international studies from the University of California, Santa Barbara (2012). She spent several months conducting research in Rio de Janeiro, Brazil, and has traveled throughout Latin America. Her current research interests lie at the intersection of urban growth and human-environment interaction as they relate to global issues of inequality, health, and development.

Rafael Guerreiro Osorio is a researcher and director of the Division of Social Studies and Policies at the Institute for Applied Economic Research (IPEA). He obtained a PhD in sociology at the University of Brasília (2009), and most of his research is on social stratification and mobility. His most recent publications are "Privatization and Renationalization: What Went Wrong in Bolivia's Water Sector?" (with Degol Hailu and Raquel Tsukada, *World Development*, 40, 12, 2012) and "The Persistence of Black-White Income Differentials in Brazil," in *Affirmative Action in Plural Societies* (Palgrave Macmillan, 2012).

Jan Nederveen Pieterse is Mellichamp Professor of Global Studies and Sociology at University of California, Santa Barbara, and specializes in globalization, development, and cultural studies. He is the author of ten books and (co)editor of eleven books. His URL is http://www.jannederveenpieterse.com.

Gustavo Lins Ribeiro has a PhD in anthropology (City University of New York, 1988). He is a full professor of anthropology at the University of Brasília and a research fellow of Brazil's National Council of Scientific and Technological Development (CNPq). He has written and edited fourteen volumes in Portuguese, Spanish, and English, and more than 180 chapters and articles in different countries. He is a former chair of the Brazilian Association of Anthropology and of the World Council of Anthropological Associations. He is a vice president of the International Union of Anthropological and Ethnological Sciences and chair of the National Association of Graduate and Research Centers in the Social Sciences (Brazil).

Livio Sansone (Palermo, Italy, 1956) received his BA in sociology at the University of Rome and MA and PhD at the University of Amsterdam (1992). Since 1992 he has lived in Brazil where he is an associate professor of anthropology at the Federal University of Bahia (UFBA). He is the head of the Factory of Ideas Program—an advanced international course in ethnic and racial studies—and coordinates the Digital Museum of African and Afro-Brazilian Heritage (http://www.museuafrodigital.ufba.br). He has published extensively on youth culture, ethnicity, inequalities, and globalization with research in the UK, the Netherlands, Suriname, Brazil, and, recently, Cape Verde, Guinea Bissau, and Senegal.

Myrian Sepúlveda dos Santos is an associate professor of sociology at the State University of Rio de Janeiro. She leads the research group "Arte, Cultura e Poder" (http://www.artecultpoder.org) and the Afrodigital Museum—Rio (http://www.museuafrodigitalrio.org). She has a PhD in sociology (New School for Social Research), an MA in sociology (IUPERJ), and a BA in history (UFF). Her research interests relate to collective memory and social and cultural theory. Her publications include several books and articles on social theory, museum exhibits, popular culture, carnival festivities, race, ethnicity, and, more recently, prisons and trauma.

Ilse Scherer-Warren is a full professor of sociology and political science at Federal University of Santa Catarina and coordinator of the Nucleus of Research on Social Movements, UFSC. Previously, she worked at the Federal University of Rio de Janeiro and spent sabbaticals at the University of London, University of São Paulo, University of Brasília, and University of Minas Gerais. Her fields of interest are social movements, civil society, social networks, citizens' rights, social inclusion, and affirmative action. Among her recent publications are the books *Redes emancipatórias: nas lutas contra a exclusão e por direitos humanos* (Curitiba: Ed. Appris, 2012) and *Movimentos sociais e participação: abordagens e experiências no Brasil e na América Latina* (with Ligia Lüchmann) (Florianópolis: Ed. UFSC, 2011).

Sergei Soares is an economist whose main fields of study are discrimination, poverty, inequality, cash transfers, and education. He is currently a researcher at the Institute of Applied Economic Research and has worked extensively with the Bolsa Família program.

Pedro Ferreira de Souza is a researcher at the Brazilian Institute for Applied Economic Research (Ipea) and a PhD candidate in sociology at the University of Brasília. His research has focused on income inequality, poverty, and social mobility.

Index